Land Surveying and Levelling

PRINTED BY
LOVE AND WYMAN, LIMITED,
GREAT QUEEN STREET, W.C.

"The Builder" Student's Series.

LAND SURVEYING

AND

LEVELLING.

BY

ARTHUR THOMAS WALMISLEY,

Member of The Institution of Civil Engineers, Fellow of The Surveyors' Institution, Fellow of King's College, London, and Honorary Associate of The Royal Institute of British Architects.

LONDON:
D. FOURDRINIER, "Builder" Office, Catherine Street, W.C.
ALSO OF
WHITTAKER & CO., Paternoster Square, E.C.
1900

LAND SURVEYING

AND

LEVELLING.

PREFACE.

IN the Author's opinion, no existing treatise upon the subject of Land Surveying is complete as a single text-book which will enable a student to become self-taught, nor does this book aim at such distinction. Efforts have been made by others to render accessible in a single volume what heretofore was scattered throughout many; but the criticisms of reviews show the disappointment of such attempts. A few lessons in the field received from an experienced surveyor are indispensable for proficiency, and will provide an opportunity for valuable suggestions to the student as to what he should read up carefully and what he may skim over. In the acquisition of such book-knowledge, he is recommended to study four or five treatises by different Authors, and to exercise his own judgment, aided by the information received from his Instructor as to the methods he adopts. Throughout the compilation of both this and its companion volume, "Field Work and Instruments," one purpose has been persistently kept in view, namely, the practical aim of the work. Every surveyor knows that much time is saved in the execution of work, by the due exercise of forethought, but without knowledge, forethought is practically impossible. Great patience must be exercised by the young beginner He must remember that it is better to be blamed by his employer for being slow than for being inaccurate, and he will then find, with due exercise of care and precision that he will acquire speed in proportion as he is diligent in practice.

Efforts have been made to supply the needs of architects, engineers, and surveyors in the following pages. Railway Surveying and Parliamentary work have been dealt with, in special chapters, as useful for engineering students. Architectural students must not, however, conclude that this book needs no place on their shelf. Part of architecture consists in the consideration of a site with a view to placing a building on the land, and in the case of a mansion to be built on a new estate, an architect who surveys the estate himself, will know much more about the ground, than an architect, who, without inspecting the site has employed a surveyor to do the survey for him. The laying out of an estate round a mansion, needs to be considered in relation to the mansion itself, as the centre of the whole.

Precautions in the handling of instruments are given, but the limits of space allow their actual use to be only briefly described. Once, however, the construction of an instrument is understood, and the general principles of its application become realised, a little out-door practice will contribute more towards efficiency, than several hours spent in reading written descriptions of field-work.

The Author is indebted to several well-known instrument makers, as well as to previously-published Student's Column articles in THE BUILDER, for the illustrations which accompany this work.

A. T. W.

August, 1900.

CONTENTS.

CHAPTER		PAGE
I.—INTRODUCTION		1
II.—RANGING OUT A LINE		4
III.—THE CHAIN AND CHAINING		11
IV.—SETTING OUT PERPENDICULARS		23
V.—USE OF TRIGONOMETRY		35
VI.—FIELD-BOOK		46
VII.—POSITION OF BASE LINES		53
VIII.—PLOTTING ANGLES		73
IX.—TRAVERSE SURVEY		84
X.—PLOTTING A PLAN		90
XI.—COMPLETING AN ESTATE PLAN		111
XII.—ORDNANCE MAPS		127
XIII.—ENLARGING AND REDUCING PLANS		141
XIV.—COMPUTING SCALES AND PLANIMETERS		149
XV.—TAKING LEVELS		163
XVI.—LONGITUDINAL AND TRANSVERSE SECTIONS		180
XVII.—CONTOURS		191
XVIII.—SETTING OUT CURVES		203
XIX.—MARINE SURVEYING		226
XX.—PLANS FOR DEPOSIT		234
XXI.—PARLIAMENTARY SURVEYING		239
XXII.—RAILWAY WORK		309
SCHEDULE SHOWING THE PROFESSIONAL PRACTICE AND CHARGES OF ARCHITECTS		325

LAND SURVEYING AND LEVELLING.

CHAPTER I.

INTRODUCTION.

The origin of Land Surveying has been disputed by historians, but it is generally supposed to have originated with the Egyptians, who were compelled by the annual inundation of the River Nile, removing or defacing the boundaries of different proprietors, to devise some method of ascertaining the position of their land-marks by measurement after the waters had subsided. From the banks of the Nile, it is said to have been introduced into Greece by Thales, who was born 640 B.C., after he had travelled in Egypt with a view to study mathematical science, and to him as well as to Euclid, Archimedes and others, we are indebted for many geometrical deductions which have proved of great value in improving the art of measuring land. So great was the value set by the ancient Romans upon a knowledge of the art, that they accounted no man capable of commanding a legion who was incapable of measuring a field.

The area of permanent pasture in Great Britain is becoming largely increased, and dependence upon wheat supplies from abroad is becoming more and more absolute, because wheat cannot be profitably grown here, and because the profits of barley growing are so uncertain. Towns extend in all directions giving an increasing value to land, with the consequent necessity to ascertain accurately its dimensions and content.

Compendiums of mensuration, together with trigonometrical expressions for the calculation of heights and distances, form no small portion of existing treatises upon the subject; but a knowledge of the practical use of instruments is of primary importance, and in this study, it is essential to start upon true principles, otherwise it is difficult afterwards, to steer clear of error.

Professor Leslie, in his Geometry, writes: "In surveying hilly ground it is not the absolute surface that is measured, but the diminished quantity, which would result, had the whole been reduced to a horizontal plane, the distinction being founded on the obvious principle that since plants shoot up vertically, the vegetable produce of a swelling eminence can never exceed what would have grown from its levelled base. All the sloping or hypothenusal distances are, therefore, invariably reduced to their horizontal lengths, before the calculation is begun."

Thus we see the practice of professional surveyors supported by the experience of a leading mathematician of his day, and the same opinion is approved by all who have had to deal with the value of land, for it is evident that if a field be sold by the surface measurement, and if it be afterwards levelled by filling up the valleys with material from the hills, the owner would not have the same quantity of land as he had paid for, if the field be re-measured. The only thing that can be reasonably admitted is the work of the labourer, consisting wholly of lineal and superficial measurement, such as mowing, reaping, hedging, ditching, etc. Lastly, no more houses could be built upon sloping ground, than could be erected upon its base, if the site were levelled by the removal of the hill. Hence the purchaser of ground, situated on a hill, would be unjustly treated, if the area be calculated by hypothenusal dimensions.

In the field, a surveyor has, in the first place to consider what lines are necessary to be measured, not for the purpose of obtaining the area, but from which to measure the necessary offsets, in order to secure a correct outline plan of all the short indentations contained within the boundaries, and to accurately connect together the whole survey. The area can then be expeditiously calculated from the plan. It is a mistake to rush on to the ground in order to commence a survey before you have carefully examined the extent, shape, and position of all you have to deal with. A little time spent in first walking over the ground will save time in the long run. The scale to which the plan is required to be plotted should be also settled at the outset, if possible, as upon this decision would depend the amount of measurements needed to be taken for the offsets.

The importance of being provided with a change of clothing to suit all weathers cannot be overrated. The only real impediment a surveyor has to accept is a thick fog. Good stout boots are essential in field work. The comfort of shooting boots, where much walking about on rough ground has to be done, is only realised by those who have tried them. Two pairs are advisable to wear in turn upon alternate days. The boots are preferably of porpoise hide, well made, of good fit, free from soft lining (which is liable to pucker), broad at the tread and stout in the sole, and fastened by leather laces. Some surveyors wear breeches and gaiters. Knickerbockers, with box-cloth bands fastened with three or four buttons at the side, are better than straps or garters, which tend to prevent a proper circulation of the blood. When leggings are put over thick stockings they keep the calf of the leg too warm, and consequently encourage a feeling of laziness. Trousers are better in all weathers, as socks can be worn to suit the time of year, and leggings over the trousers in wet weather. A light mackintosh should be available. It is a good practice to establish a daily roll-call before starting in the field, enumerating the articles required for the day's work, in order that nothing may be forgotten. As regards men, it is a mistake to work short-handed. Men are needed to carry pegs and act as labourers in addition to those engaged as chainmen. In the country, it will generally be found that the "boots" at the hotel will possibly know of honest, sober men acquainted with the district, and perhaps those who have been out before as chainmen, and who will not therefore need the amount of training required by those who are not accustomed to the work.

CHAPTER II.

RANGING OUT A LINE.

In making a survey the first thing to be done is to determine the best positions for running the straight lines upon which the survey is to be based, and the next is to range out the base lines before measuring them from one station to the other. Ranging rods, of lengths varying from 5 ft. to 7 ft., are usually made circular in section, and painted in lengths of one link (7·92 in.), 9 in., or 1 ft., white, red, and black alternately. When one colour at a distance cannot be clearly seen against a background, one of the other coloured portions can generally be distinctly observed. White is most suitable for most backgrounds, but for a pale grey aspect, such as is presented by ploughed land, black is preferable, and for a green background, a red colour on the rod is preferred. For use at a considerable distance it is well to attach a flag of red and white bunting of about the size of a pocket handkerchief permanently to the top of certain poles, or in the absence of a more suitable signal a pocket handkerchief may be temporarily employed for the purpose. Some think it useful for long distances to fasten a piece of white canvas by tapes halfway up the rod.

The surveyor must agree with his assistant as to the signs he intends to adopt for directing him in the field when setting a pole or ranging rod in a required line; and the pole must not only be fixed in the ground upon this line, but be left vertically in position. The latter precaution can generally be accomplished by eye, but where great accuracy is required the aid of a plumb-bob, as shown in fig. 2, must be introduced, the string from which the plumb-bob is suspended being turned over the first and second fingers of the hand, so that when the string hangs vertically the pole may be set parallel to it. The detail of the plumb-bob in fig. 5

(pages 6, 7) shows the top unscrewed for the purpose of attaching the suspending cord by means of a knot. Whipcord is the best to employ, as it wears better than string. A good form of plumb-bob for carrying in the pocket is shown in fig. 3, with a reversible point which screws up inside the hollow case forming the body-piece of the plumb-bob. Sometimes a square distance plate equal in width to the diameter of the cylindrical case is added, which is useful when checking the perpendicularity of walls. There is no fixed distance between the ranging rods. They serve to give direction only, except at the intersection points, when their position is accurately recorded. (See fig. 1, pages 6, 7.)

A straight line may be set out by successively placing poles in the ground in line with one another at convenient distances apart in the order marked 1, 2, 3, 4, 5, in fig. 1 (pages 6, 7), when one station B is fixed, the line being found to pass through A. The surveyor stands at a short distance from A behind the pole looking towards B with both hands free. The accuracy of the operation is promoted by not letting the eye be too near the pole A from whence the observation is made. If the intermediate pole requires to be moved towards his left, he indicates this to his assistant by moving his left hand in that direction. If the intermediate pole requires to be moved towards his right, he directs his assistant with his right hand in a similar manner. The assistant places himself a little off the line, facing a direction at right angles to the line, so that he does not impede the line of sight and can turn his head round to receive further instructions from the surveyor. He must only hold the pole which he is directed to set with one hand and the surveyor in giving him the signals for so doing, should present the palm of the hand he holds up towards his assistant, as the full width of the hand is better to see than the edge. Any other spare poles should be laid down upon the ground, or be carried by another man, or boy, at a short distance from the line, otherwise they may confuse the surveyor in setting the pole required. Should the bottom of the pole be in the correct position, but the pole itself not quite upright, the surveyor may hold the elbow of his right arm with his left hand and move the right hand, palm

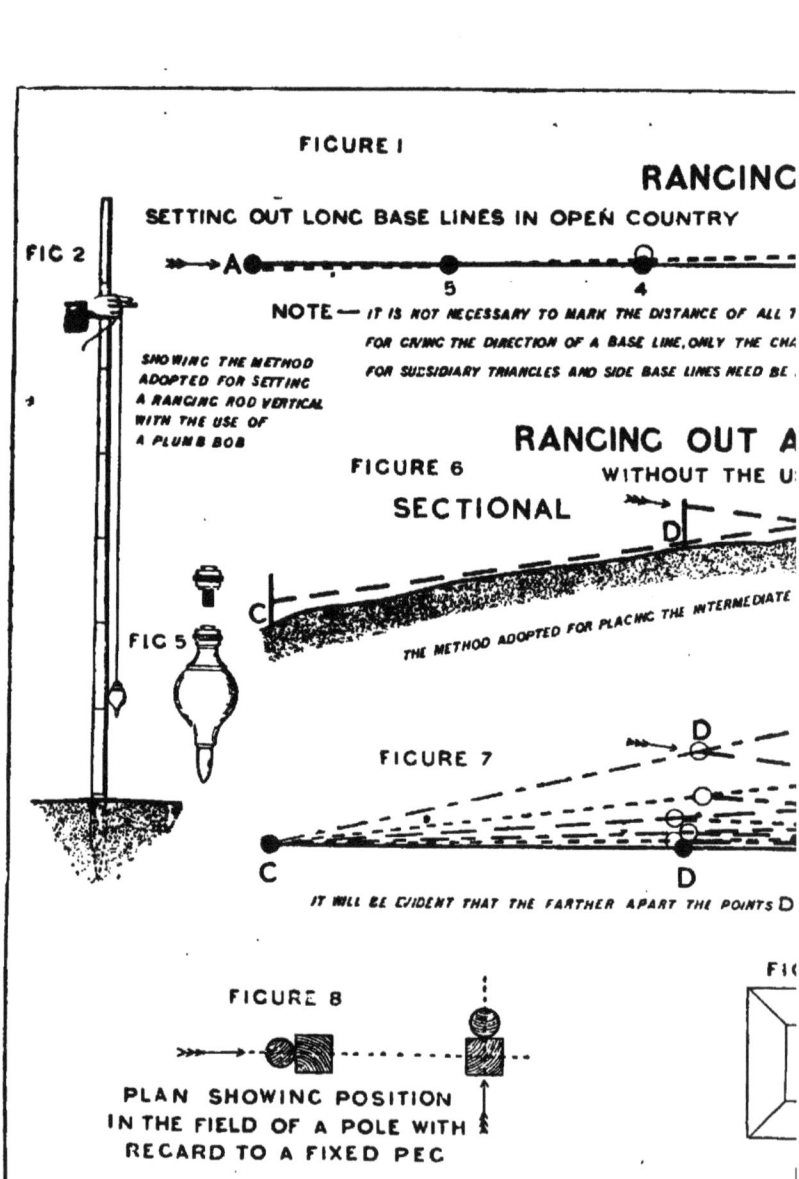

Land Surveying and Levelling, pp. 6, 7.

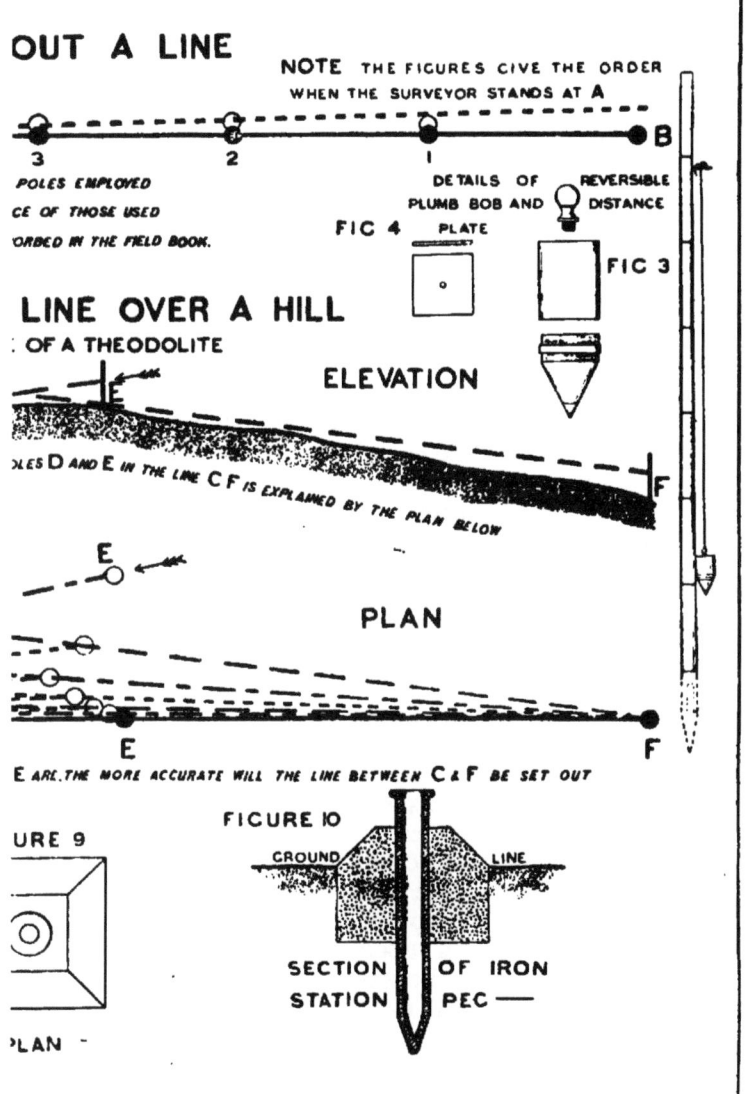

outwards, to the right, like an inverted pendulum, or else he may hold the elbow of his left arm with his right hand and move the left hand palm outwards to the left. By such means he can readily indicate to the assistant that the bottom of the pole is correctly placed, but that the top requires to be moved more, either from or towards the assistant, in order to appear upright. If the line to be set out be a very long one, the surveyor may render his signals intelligible at a distance to his assistant by waving a pocket handkerchief in the direction required by either hand, or by fastening a pocket handkerchief to the end of a spare ranging rod and employing this as a flag to indicate his instructions.

When any intermediate pole is set accurately, both as regards position and uprightness, the surveyor indicates that it is correct by raising both hands, palm down, one upon each side of the pole at A, and, while slightly bending his body, moves them from about the level of his head towards the ground. The assistant then presses the pole into the ground with both hands, or, if the ground be hard, props it up with loose stones or other material, and looks towards the surveyor for final instructions as to leaving it perfectly vertical before he proceeds to set the next pole. If both the stations A and B are fixed and the intermediate points upon the line have to be determined, the best order in which to place the poles in position, is to commence near the distant station and allow the assistant to work towards the surveyor along the line, as by this means each pole that is being fixed is distinctly seen by the surveyor, and is not in any way hidden behind the other poles. Where possible the surveyor should observe in each case the bottom or lower end of the ranging rod. If the pole marked 5 in fig. 1 is fixed first, the remaining poles might appear in line when a complete diameter out as indicated upon the diagram, because daylight could not be seen between them in the direction A B, and the base line A B would, if continued beyond B or if the pole B does not stand perpendicular, take the direction of the dotted line. Although this would not affect the actual length of the line measured between the points A and B, it would affect the length of lines measured off A B upon either side of it, as it is clear that if the chain be laid out of line, an offset will be measured either too short or too long, the consequences being an incorrect plan leading to inaccurate computation of area.

Figs. 6 and 7 (pages 6, 7) illustrate the ranging out of a base line between two fixed stations C and F when a hill intervenes. The poles C and F are fixed perpendicularly in the ground, and the surveyor with one of the intermediate poles D or E and his assistant with the other intermediate pole, each proceed to a distance as far apart from one another as will enable the top half of the pole at C to be viewed from E and the top half of the pole at F to be viewed from D. The diagram, for the sake of clearness, shows the poles D and E shifted about upon one side of the line C F only. It is probable that one or both of them in the actual setting of the lines, C D E and D E F would be brought too much over the line C F, and would have to be shifted back and wriggled into proper position by the process illustrated in fig. 7. It is evident that the farther apart the points D and E are from one another the more accurate will the line between C and F be ranged out, as a longer length of line will be common to the two sides C E and D F (figures 6 and 7).

When the ground is very undulating the surveyor is compelled to set up the poles at very short distances apart. When the line crosses a thickset hedge, a gap must be cut through with a bill-hook, sufficient to make the continuation of the line upon the opposite side clearly visible.

Poles are generally understood to be longer than ranging rods, but are the same in principle. Both are usually shod with a pointed iron shoe for convenience of planting in the ground. The most suitable ranging rods taper from the bottom to the top, being over an inch in diameter at the shoe, and about $\frac{3}{4}$ inch at the top. When the position of a station point upon or at the termination of a base line is marked by a wooden peg driven into the ground, it is advisable to observe the precaution illustrated in fig. 8 (page 6) in order that the pole may appear in the direction of the line viewed. In rough open moorland iron pegs fixed in concrete, as shown in figs. 9 and 10, may be advantageously employed for the main stations. The hollow portion serves to hold a ranging rod, which can be easily wedged up vertically in setting it up with the aid of a plumb-bob (see fig. 2). The top of this iron peg (fig. 10, page 7) serves as a reliable bench mark when taking the levels.

A moderately gentle slope is by no means objectionable, as it enables the surveyor to obtain a better view of the country in ranging out his lines (see also pages 57 and 72, but in railway work the precaution named upon page 309 must be observed). That the setting out of a straight line on the ground from one point to another demands some care and skill is easily tested by attempting to cross a field covered with snow, or other ground leaving footmarks, directly from one object to another. You must view two objects in line, one visually covered by the other to guide you for the continuation of the line in a straight direction. With one object only, as a guide, an irregular course will be pursued, while by means of continuously covering coincident sights, a nearer approach to a straight line is obtainable. Hence it is necessary to have at least one intermediate pole upon a line in addition to the poles at each end.

CHAPTER III.

THE CHAIN AND CHAINING.

ALL lines upon which a survey is based are measured in a perfectly straight line from end to end, for which purpose a surveyor has to select one of two kinds of chain and tape which are commonly employed in the measurement of land, each divided into 100 links. The total lengths of the best known short chain and tape are each 66 ft., and of the long chain and tape 100 ft. The former is called Gunter's chain, from its inventor, the Rev. Edmund Gunter (1620 A.D.), and its use is quite peculiar to this country. It consists of 4 poles, or 22 yards, decimally divided into chain links, with eyes at each end. Each link, being one hundredth part of 66 ft, will be equal in length to 7·92 in., but this fact is more interesting than useful, as any portion of a chain is invariably expressed in links. Chains of metre lengths, divided to fifths of metres and tallied at every two metres, are also sold for work requiring a metrical measurement.

Link Chains.—A long chain saves time over level ground in fixing arrows in the ground; but where the ground is uneven and rough, the links being so long as compared with those of a Gunter's chain, are apt to be twisted in being drawn through hedges and rough places, which of course produces errors in the measurements. In open ground a long chain aids accuracy, because the more often the whole length of a chain has to be shifted in measuring a base line the greater are the chances of error in recording the total measurement. A chain 100 ft. long is a suitable length, as a longer chain would be found very inconvenient to drag, especially in wet weather.

In folding up a chain the most expeditious plan is to roughly lay it out as in fig. 1 (pages 14, 15); then take the links nearest the 50 mark in one hand and fold the chain double, until the handles are reached, taking care so to cross the links which come next to one another in folding, that the body of the chain, when folded, may be smaller in the middle than at the ends, as shown in the figure representing the chain folded up. A strap is first passed round the middle to bind the links together and is then passed through the handles together, with the heads of the arrows, and fastened by a buckle. When you wish to unfold the chain, having unstrapped the arrows, take both handles in one hand and, having freed the first two or three links, take the remainder of the chain in the other hand and throw it out from you taking care to keep hold of the handles. When thus thrown out upon the ground you must then straighten and adjust the links where necessary and close any of the connecting rings that may be open at the joints before proceeding to measure the base line.

Every chain is accompanied by ten arrows, each about one foot in length, made of stout iron or steel wire, and pointed at the bottom. The reason why ten is the number adopted is probably that ten chains of 66 feet in length equal one furlong. They are usually bent in a circular form at the top for convenience of handling, and a piece of red cloth is often attached to the ring of each, in order that they may be more easily observed when fixed in the ground. When in use they may be best carried by attaching the buckle end of the strap, which is taken off the chain, to the ring or top of one arrow, and passing the strap through the rings of the remaining nine arrows from which they can be easily removed, one at a time, as required. The use of the arrow with the strap upon it will then indicate that the whole ten arrows have been employed in the measurement of any given base line. Some surveyors adopt the plan of using ten supplemental arrows, distinguished from the others by brass or other marks, to be put down in succession at each tenth chain. In chaining long base lines it is advisable to put down a peg at every ten chains, entering it in the field book (which will be described further on in these pages), although, as here stated and subsequently referred to, the usual practice is for the ten chains to be only temporarily located. In case, however, of an accident in dropping an arrow, it becomes doubtful what is the exact number of chains that have been measured, and the position of a peg so placed limits the possibility of an error to a distance of ten chains, and then only this portion of the base line, in which a surveyor has the misfortune to lose an arrow, need be re-measured. Time is a more important element than the cost of labour and material in the use of pegs.

For the purpose of testing the chain employed in an extensive survey during the progress of any work, which will most likely stretch a little every day it is in use, it is well to fix two pegs upon a level piece of ground near a fence, and at a distance apart just sufficient to enable the outside of the handles of a correct chain when drawn tightly to touch the inner sides of the pegs. This arrangement is better than making the chain's length measure from centre to centre of the pegs, but the chain should be pulled quite taut, and the rings cleared of all dirt, and the links straightened, so that the chain may play freely along its whole length. If the chain has been previously pulled over a ploughed field upon a wet day the rings uniting the links will have become clogged with dirt and the chain will need washing, which can be effected by passing it through a stream carefully. The test distance may be set out very accurately with a level staff, or, better still, with two level staves placed end to end in measuring the line, provided each level staff has been previously tested upon the Government standards. It is sometimes well to keep a properly-tested spare chain in reserve, to be used only for the purpose of testing when a level staff is not near to hand, and the distance should be proved prior to each testing of a chain, as pegs have been known to be purposely moved by parties interested in opposition.

No survey can be accurate in which the base lines are ranged in the least degree curved, and in chaining up and down hill by a process called "stepping," as exemplified in fig. 4 (pages 14, 15), the steepness of the slope will regulate the distance between the successive points, P, R, S, as the chain can only be raised to a certain height by hand,

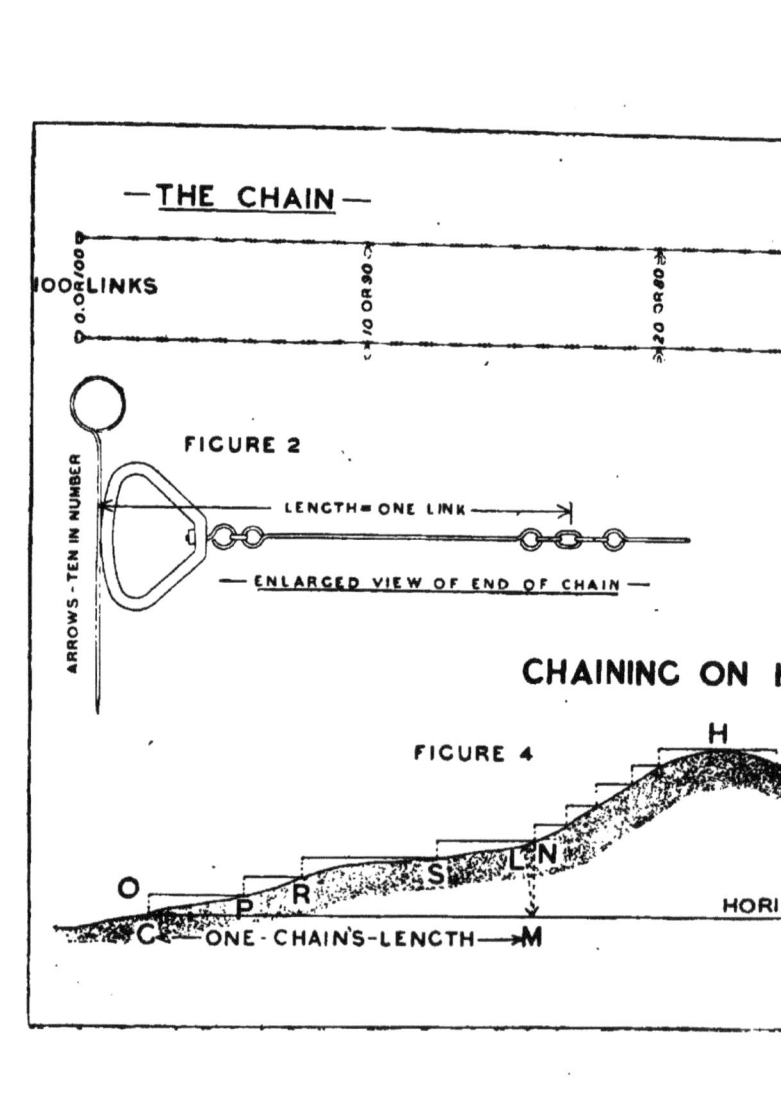

Land Surveying and Levelling, pp. 14, 15.

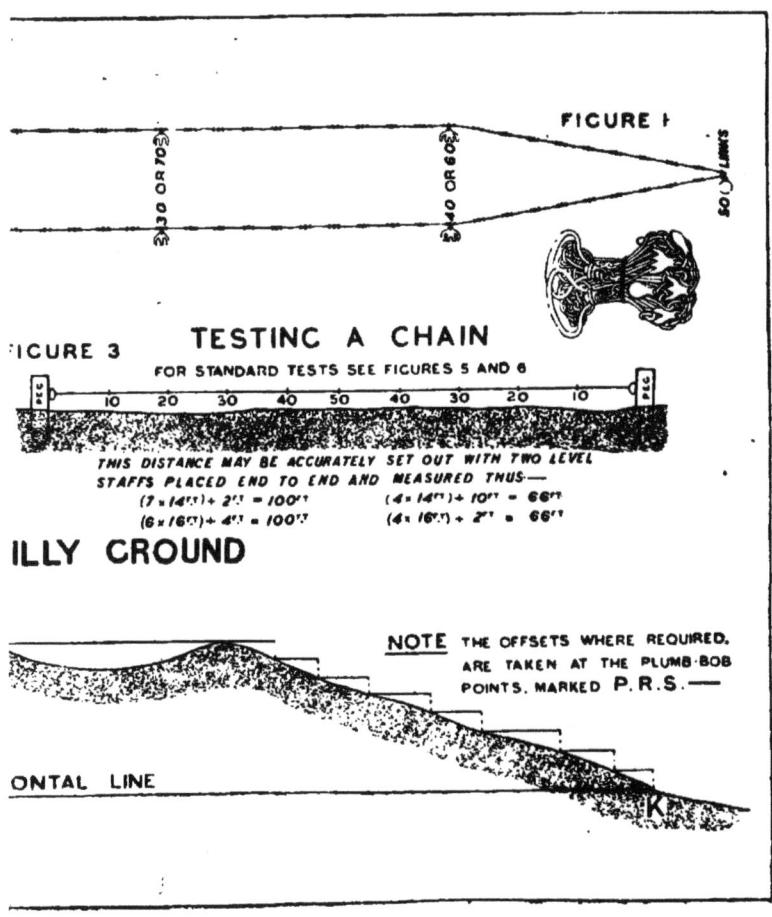

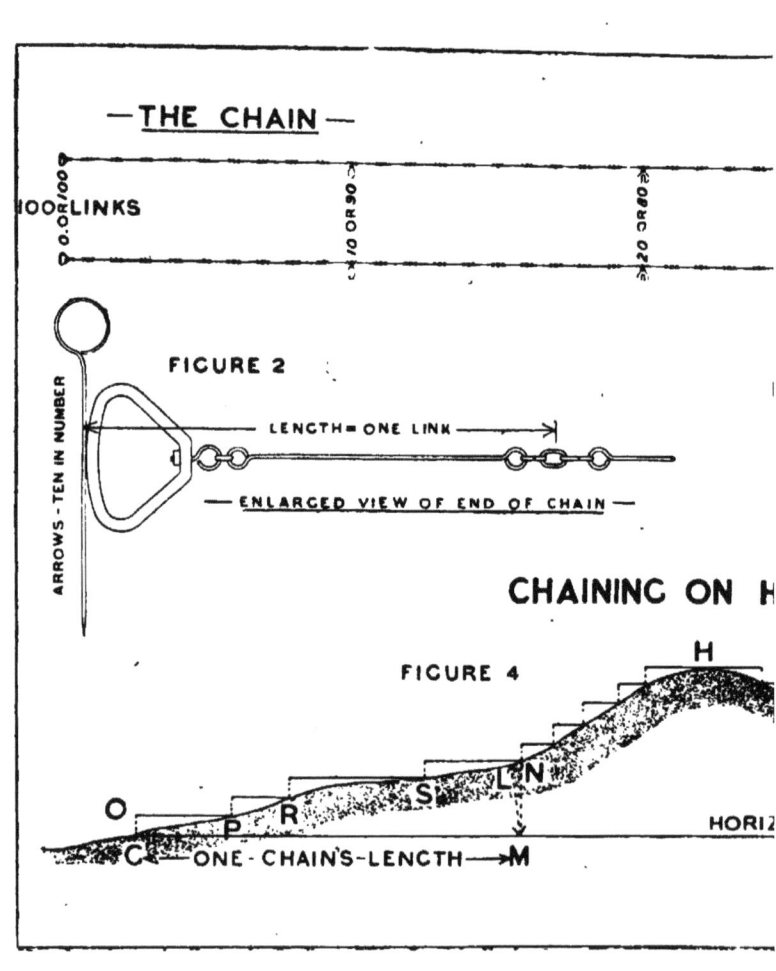

Land Surveying and Levelling, pp. 14, 15.

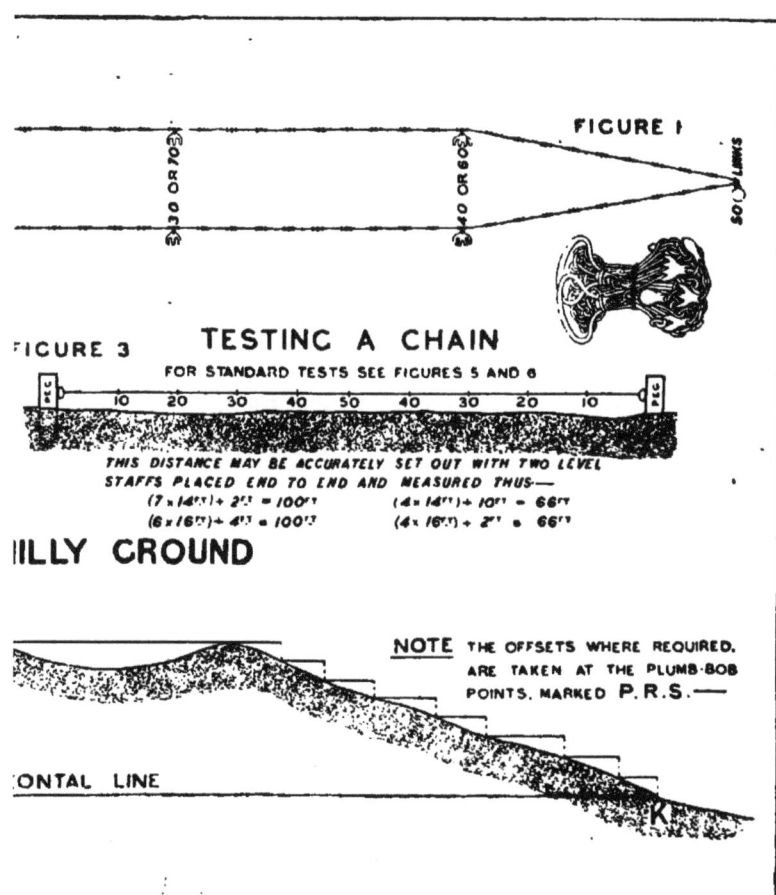

and its top end should in each sub-divided length be held close to the ground. The chain should also be well stretched between the points of sub-division. In chaining downhill a plumb-bob is superior to a dropping arrow, for the reason that the head of such an arrow is liable in falling to displace the point out of the perpendicular. A plumb-bob, however, takes longer to manipulate than a dropping arrow, although it can be adjusted with great care. In chaining uphill the follower manages the plumb-bob. In chaining downhill the leader takes charge of it. In either case, at the end of each chain's length, the totals are corrected, by the follower returning all arrows picked up, from intermediate points in a chain's length, to the leader, prior to proceeding with another chain's length.

Tapes.—It may be mentioned that there are two descriptions of linen tapes; one is usually known as the metallic tape, and has delicate copper wires or threads interwoven with the substance of which it is composed. The other kind is a plain linen tape without any such additional combination. When really good, either of them may be trusted at any time to half an inch while new, but linen tapes stretch by use in windy weather and shrink by use in rain. In using a tape in wet weather, or upon any occasion when it gets wet, it should never be rolled up until it is quite dry. Winding up a wet tape and laying it by in its box until it is next wanted, is a certain means of spoiling it. The tape, after being washed, should be coiled loosely up, and after carrying it for a short time in the open air it will be dry enough to wind up. The same remark applies to rolling up a dirty tape. In winding up a tape the box is held in one hand and the portion outside the box is drawn between the first and second fingers of that hand by turning the handle by the other hand. By this means the tape is prevented from twisting as it enters the box or leather case, which is usually of a polished brown colour. (See pages 17-19.)

Metal band tapes are now very much employed instead of chains, as the sub-divisions indicated upon a continuous band are found to remain more accurate than is possible

Land Surveying and Levelling, pp. 17, 18, 19.

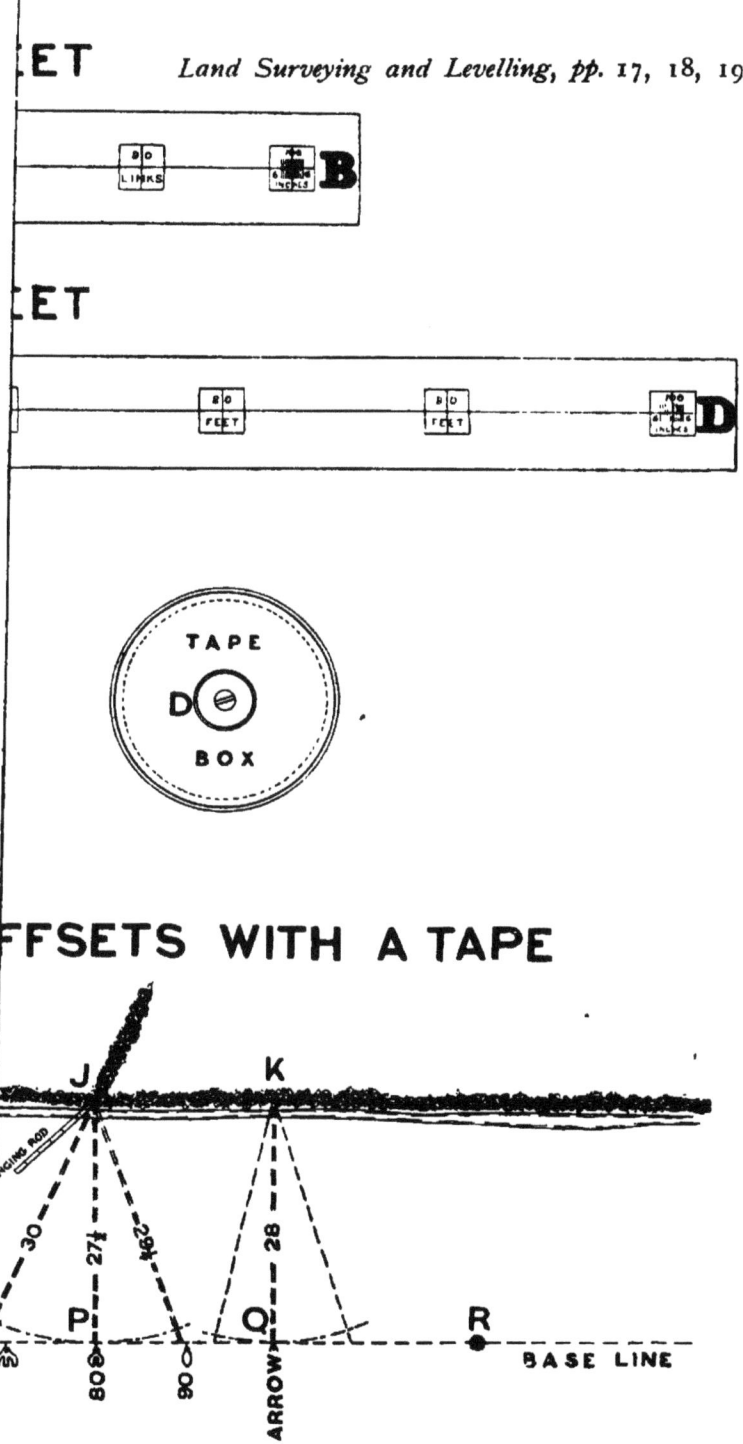

with the use of separate links. The band tape also makes a reliable standard by which to compare an ordinary link chain or a linen tape. The band chain or band tape is most conveniently wound upon an open X reel, but some are contained in a metal case.

Fig. 8 (pages 17-19) illustrates a part of the measurement of a base line, and the method of taking offsets from one side of it to a fence line. The station or point upon a line from which the measurement is to commence being first determined, the surveyor makes it his business to count the arrows before commencing to measure a line, in order to see that his leading chainman has ten, and the surveyor instructs him to put two or three of his fingers through the handle of the chain, and with his thumb to hold one of the arrows close to the outside of the handle, and at about 2 or 3 inches from the point of the arrow. Then the leader carrying all the arrows draws the chain in the direction of the line to be measured, and is guided by signs from the follower who sights the line of poles or other marks, which determine the direction of the base line, by directing the leader to bend on one side clear of the line, with the arrow exactly upright in the same hand as the handle of the chain, as described (fig. 2, page 14), to move the chain end and arrow on one side or the other, by saying "from you," "to you," and to say "down" when in line for fixing. The ten arrows are used as tallies, thus :—
When the leader reaches the full stretch of the chain, the other end being held fast by the follower, the leader holds an arrow vertically, as shown in fig. 2 (page 14), against the outer edge of the end ring or brass handle of the chain, and after shaking and pulling the chain to ensure a straight line, the leader thrusts No. 1 arrow into the ground. No surveyor will be able to record an accurate measure of length if he allows his chainman to hold long arrows at the top. He then leaves it there, and advances with the chain until the follower who has hold of the other end of the chain reaches No. 1 arrow, and calls upon him to stop. The leader places another arrow against the brass handle or terminal ring of his chain, which is again stretched and directed as before. He then fixes No. 2 arrow into the

ground and leaves it there, to show the follower what point upon the base line he has next to measure from. The follower now picks up No. 1 arrow, and both the leader and the follower advance as before, until the required length of line is measured, or until all the ten arrows have changed hands, when a pole can be temporarily placed in line to mark the position of the tenth arrow, or length of one thousand links, while the follower hands all the ten arrows to the leader. The peg, recommended on page 13, can then be fixed.

A good chain leader will whenever possible view some back object in the direction of the chain line, so that he may be able to keep the chain in a straight line with very little correction by the chain follower. If the ground is too hard to press the arrow at the end of any chain's length into its place, the leader marks the ground with the point of an arrow, thus, $\overline{\wedge}$ and lays the arrow down. The intersection of the lines in this mark shows the follower to which point the end of the chain has to be held in continuing the measurement of the line. While necessary for the leader and follower when chaining to stoop in a bending attitude to the arrow in the ground, there is no occasion for either of them to go down upon their knees to the ground.

In taking offsets, the surveyor reads the tape at the chain, and the ring of the tape being held at a point to which the offset is required, the surveyor twists the tape round in a horizontal plane, as shown in fig. 8 (pages 17-19), in order to read the shortest measurements, and also to ascertain the link upon the chain at which the offset would form a perpendicular to the base line.

When the points to which offsets are to be measured happen to be situated near the chain line, no difficulty is experienced in taking the measurement at right angles to the base line. For lengths exceeding the number of links which can be plotted by an ordinary offset scale, it is advisable to introduce an instrument such as a cross staff or optical square, rather than estimate the direction of a right angle by eye. (See "Field Work and Instruments.")

When a ditch intervenes, the centre of a hedge may be reached by placing the spiked end of a ranging rod in the ring of the tape C, as indicated in fig. 7, and proceeding as in fig. 8. If the surveyor stands in a field with the ditch on his side of the hedge, he knows that the field in which he stands reaches, as a rule, only up to the edge of that ditch. Sometimes there exists what is termed a footset hedge, that is, having no ditch either side. The offset should then be taken to the centre, and should be noted in the field book thus ——⊥——. It should also be observed that where a fence is found or hedge growing situated upon a bank having a ditch upon one side, it is probable that the bank has been made out of material originally dug out of the land of the owners of the fence in order to form the ditch, and in such case the boundary would be, say, 4 ft. more or less from the centre of the fence, according to the practice of the locality, so as to include the ditch, or the brow of the ditch may be the boundary. The surveyor should make local inquiry in order to arrive at a correct conclusion. If an old map of the property exists it should certainly be inspected, as many doubtful points upon the question of boundaries may be settled by it. (See also pages 135 to 137.)

An offset staff is useful for hedges, where it may be difficult otherwise to reach the centre. It consists of a flat rod about 1½ in. wide by 1 in. thick, and generally 10 links long, divided into links. When a tape is employed, the surveyor must take care that his assistant has the end of the tape at the point to which the offset is to be taken, that he holds the ring without holding 5 ft. or 6 ft. of the tape in his hand, and that the tape is not short through being sewn in any part of its length. Where offsets are required upon a slope, they should be taken as much as possible at or near the plumb-bob points, so as to be accurately measured off the base line. Should the base line to be measured be shorter than ten chains, the number of arrows in the hand of the follower is counted and the odd links added.

CHAPTER IV.

SETTING OUT PERPENDICULARS.

A PERPENDICULAR in a horizontal plane to a base line may be set out with the use of a chain only, in the manner shown by figures 1, 2, and 3 (page 24). A right-angled triangle is formed, having its sides in the proportion of one set of numbers given in the accompanying table to fig. 1; but the proportion of 3, 4, and 5 is generally selected, as the points of intersection at the angles are here more clearly defined than they would be if the proportion of the other lengths were taken. The proportions are based upon Euclid, book I., prop. 47, a theorem said to have been discovered by Pythagoras, a disciple of Thales, who, after travelling in India and Egypt in pursuit of knowledge, settled in Tarentum, in Italy, where he founded the celebrated Pythagorean school, 550 years B.C. A distance of 40 links is measured along the base line in fig. 1 from the point at which the direction of a right angle is required. The assistants then hold the handle of the chain at V, and the 80th tally at U, while the surveyor holds the 50th tally, and pulls up the chain at T, so as to form the sides of the triangle T U V, respectively equal to 50 and 30 links with the chain. T U is thus set out at right angles to V U. If, instead of 40 links, 30 links be measured along the base line from U to V, then the 90th tally must be held at U to obtain a similar triangle as shown in fig. 2. Should the chain be suspected to be out of order, giving a result as indicated by the dotted lines in fig. 2, the exact direction of a right angle may be obtained by setting out the triangle upon both sides of a perpendicular and bisecting the distance marked by the 50th tally upon the chain to obtain the position of the full line, shown in fig. 2, between the dotted triangles. If the 50th tally comes to the same point in both cases it

SETTING OUT PERPENDICULARS.

SETTING OUT A PERPENDICULAR WITH THE CHAIN ONLY

$$TU^2 + UV^2 = TV^2$$
$$3^2 + 4^2 = 5^2 \text{ (EUCLID·BOOK I·PROP)}$$

PERPENDICULAR SIDES		INCLINED SIDE
3	4	5
5	12	13
7	24	25
8	15	17
20	21	29

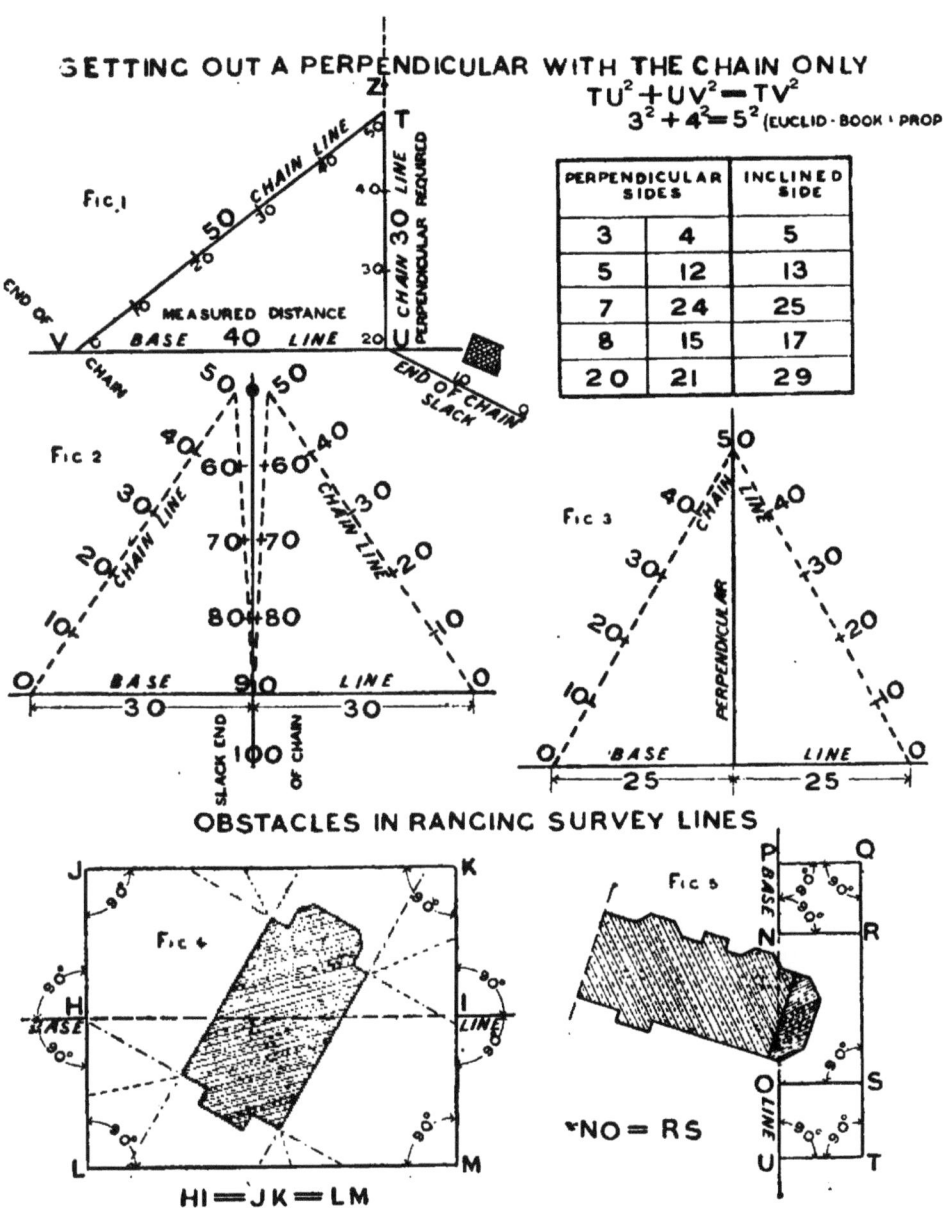

OBSTACLES IN RANGING SURVEY LINES

HI = JK = LM

NO = RS

shows the method to have been correctly applied and the chain to be sufficiently accurate for the purpose. In fig. 3 an equilateral triangle is formed by measuring 25 links along the base line upon each side of the point at which the right angle is required, and while the handle at each end of the chain is held by the assistants at the points so arrived at, the surveyor pulls out the chain and marks the position of the 50th tally, through which the perpendicular line would pass. However desirable it may be to range base lines free of all obstacles, it is in many instances impossible to do so. Some impediment might intervene in a long base line which may not seem an obstacle, until a near approach in setting out a line, and if the base line forms the side of a well-conditioned triangle, it would be clearly inexpedient to divert it in consequence of an obstacle lying on its path.

Fig. 4 (page 24) shows how the distance H I may be measured when a rectangle can be formed upon both sides of it, and fig. 5 shows how the distance N O may be measured when it can be passed upon one side only. With a view of proving the accuracy of the work the perpendiculars at each end are erected upon both sides of the base line in fig. 4, while in fig. 5 double perpendiculars are set out at each end upon the side of the building that admits of this method. The buildings can be surveyed by offset measurements from the sides of these rectangles (page 24). The right angles can be best set out either by the use of the optical square or cross staff. A theodolite being adjusted generally for long vision only, will not be found to be so accurate for the short distances at which the angles are here fixed, as when longer lines are required to be ranged by its aid. (See page 30.)

It is impossible to carry out a survey having any pretensions to size without having to use (so called) false stations, in order to range round and measure obstructions, such as the intervention of a house, a pond, or a river. In the above cases (page 24) the continuation of both ranging and measurement is impeded, and although the actual deviation of the base line in these situations might not exceed a few feet upon one side or the other, yet when this occurs in the middle of a long base line, such shifting would in all

SETTING OUT PERPENDICULARS.

OBSTACLES IN MEASURING SURVEY LINES

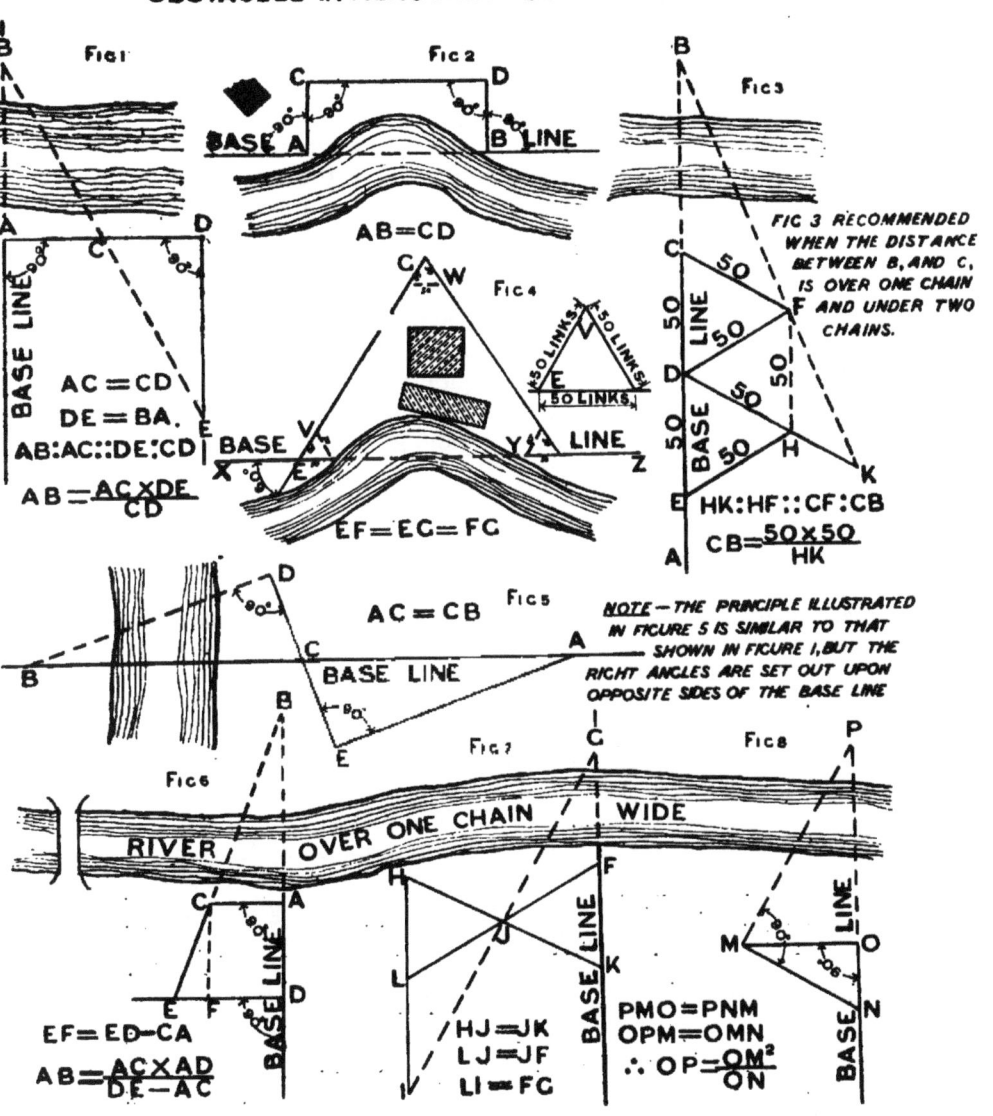

probability render it useless, as a main line of a system of well-conditioned triangulation, proper to be adopted under the circumstances of site. In measuring enclosures separated by walls, it must be observed that the ground upon which a wall stands must be included with the field to which the fence belongs, and as walls are usually broader at their base than at the top, it is important to attend to this circumstance and to note the thickness of the wall. As regards party walls, the reader is referred to the references thereto in the "Transactions" of The Surveyors' Institution. In the case of a boarded fence supported by posts and rails, where there is no ditch, the fence is usually supposed to belong to the side where the nails are driven home, presenting a fair face to the road or to a neighbour's property. Every stile and footpath (even if ploughed up) crossing a fence or boundary should be indicated in making a survey, but not necessarily every gate or opening, unless some right of way is to be denoted thereby.

Inaccessible points may generally be determined by the application of practical geometry with the use of a chain on the ground. Distances may be arrived at, without the use of an optical instrument, by measurement upon lines adjacent to the required base line, when set out so as to form any one of the geometrical constructions shown in figs. 1 to 8 (page 26). Although in figs. 1, 2, 5, 6, and 8, the right angles can be readily set out with the aid of a cross staff or an optical square, the perpendicular lines may also be set out by means of the chain only, as already explained, while in fig. 7 the system of transversals is shown. In fig. 1 the measured distance, D E, is equal to the required distance, A B. In fig. 2 the measured distance, C D, is equal to the required distance, A B, and this method would be equally applicable if the edge of a marsh or gorse had to be passed. In fig. 3 the only distance required to be measured is H K, and this can be best done with a tape. The equilateral triangles are set out in fig. 3 in the same way as fig. 4, by fixing each end of the chain at two points upon the base line, 50 links apart, and then pulling out the chain tight by means of the middle brass indicator to determine the third point in the triangle. If the box

Land Surveying and Levelling, pp. 28, 29, 30.

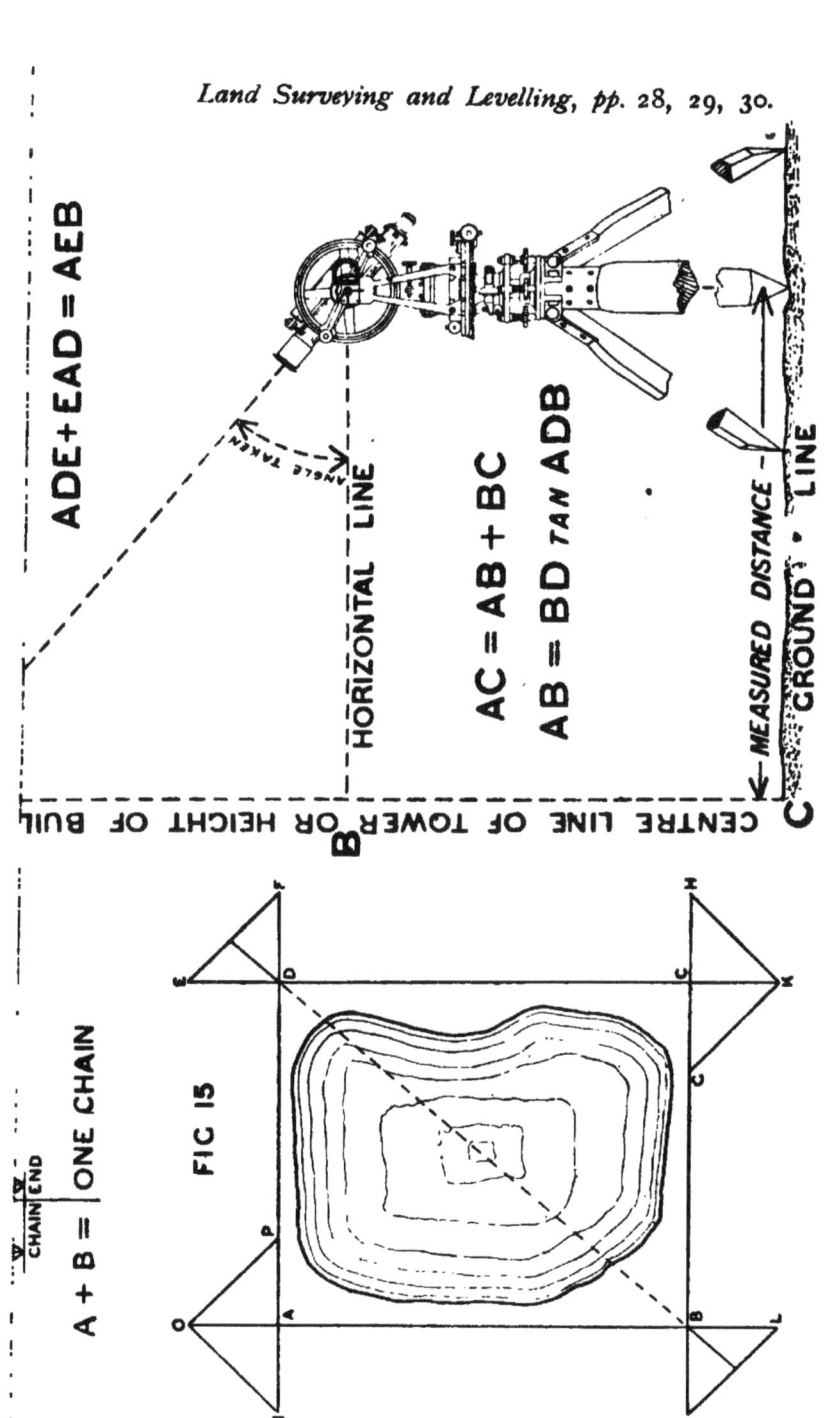

ADE + EAD = AEB

AC = AB + BC
AB = BD TAN ADB

A + B = ONE CHAIN

FIG 15

SETTING OUT PERPENDICULARS.

sextant be employed for fig. 4, it will be well to check the accuracy by erecting a pole in the continuation of the line C E at 60° with the line E X upon the other side of the base line. The construction of the remaining figures is explained by the diagram. In each case the direction is verified by simple ranging, as (figs 1 to 8) the obstruction does not impede the sight or ranging of the line, but only its measurement; and the plan of arriving at the width of crossing is equally reliable, when the base line crosses on the skew or in an oblique direction with the centre line of the stream. In fig. 6, if we imagine that, at the point A, the angle C A D be first set out as a right angle or thereabouts, and afterwards the instrument be set to half the angle C A D, then if the surveyor proceeds along the line A C until the object B subtends with A, the half angle so set, the distance A B will be equal to the distance A C, because the exterior angle C A D of the triangle C B A is equal to its two interior and opposite angles, and these being by the method adopted equal to one another, the sides subtending the angles are equal.

Fig. 9 (pages 28-30) shows how the angle of a building may be correctly determined with reference to the position of a base line, when that building is so situated that it is only possible to connect it by one line A C to the main base lines from which an estate is being surveyed. The distance of the point B along the base line being fixed, measure the lengths B D, D C, and C B, so as to be able to plot the triangle B D C, and measure the lengths C E and B E, so as to be able to plot the triangle B C E. Then, if the face of the building, D E, be straight, the line C E, when plotted, should appear as a continuation of the line D C.

Fig. 10 (pages 28-30) shows a method for connecting a base line for an off-road with a main base line. A pole is placed in the middle of the off-road, at E, so as to equalise the offsets upon either side, and another pole higher up the road, not shown in the diagram, but also placed in the centre of this road. A pole is then fixed at C, so as to be in line with the two poles fixed upon the off base line and also upon a point in the main base line B A. The direction of the line C E is determined for purposes of plotting by measuring the sides of the two triangles B E C and C E K, the points B and K being taken upon the main base line B A in such positions that the tie-lines B E and E K touch the corners F and H of the adjacent buildings. Then, by recording the distance of F from B, when measuring B E, and the distance of H from K, when measuring E K, a valuable check upon the position of these corner boundaries is obtained. They are also surveyed in the ordinary way, when taking the offsets measured from both the lines B A and C E. The same method is adopted for fixing the direction of P N in fig. 11, but in fig. 11 the triangles do not extend to the points F and H as in fig. 10. When a piece of open ground exists at one corner, as shown in fig. 10, the lines B A and C E may be connected by the formation of a triangle, C D L, in which C M is measured as a *proof* line. In the triangle B E K the line E C forms the proof line. Upon a hard road, when frequent traffic might upset the position of poles temporarily set up in the centre of the road, they may be fixed very often in gully-grates and wedged up. A chalk-mark across the grate will enable the position of any pole to be easily redetermined if disturbed by passing traffic. This method has been supposed to be adopted in fixing the base lines O S and S W in fig. 11.

Fig. 12 (pages 28-30) shows a method which can be adopted for determining the angle between the sides B E and B F of an old building supposed to be out of the square in plan. The direction of the side B F is first produced by fixing a pole at A, and the direction of the line B E, by fixing a pole at C. The measurement of the line A C fixes the angle A B C. A similar plan could be adopted in the case illustrated by fig. 15 for connecting the base lines A D, D C, C B, and B A. At A and C the triangles have been formed in the same way as the triangle N O Q in fig. 11. At B and D the tie lines in the triangles M B L and E D F (fig. 15) are ranged as a check in the same straight line. Figs. 16 and 17 show a method of connecting base lines by chain measurements when the angle formed by the base lines at their intersection is very obtuse. Fig. 18 shows the method adopted for calculating the unmeasured length B E of a base line, B D, after measuring a portion, E D, of this line, and taking the angles at the extremities of the measured portion by viewing a distant object A. Fig. 19 shows a means of arriving at the height of a tower, A C, by taking an angle at D, and measuring the distance equal to B D. If an obstruction prevents the distance B D (fig. 19) being accurately measured, the formula given in fig. 18 may be used, B E in fig. 18 being supposed to correspond with B D in fig. 19. (See pages 28-30.)

Various other cases will arise in practice, each of which must be dealt with according to its special circumstances. All we can do is to suggest examples which we think to be of the most frequent occurrence, but we trust the study of the cases we have illustrated may help the student to form a correct judgment in applying main principles to such cases as may come under his care. In laying out new fences, the Surveyor should avoid the necessity of having to convey any surface water away by underground means if possible, and should a fence be required to divide arable and pasture land, it is preferable to place the ditch in the arable field, as cattle in a pasture field wander towards the ditch, and damage the boundary. All surface water should be carried away by furrows, ditches, and open watercourses, in order to limit the provision of pipes to purposes of underground drainage as much as possible.

When the theodolite is used for continuing a base line through a station marked by a wood peg, or for taking an angle between two base lines which intersect at a station marked by a peg, the best way to secure accuracy is to set up the instrument with the plumb-bob hanging vertically over the hole in which the pole originally used for ranging the line has been placed at the side of the peg and not over the peg itself. Occasionally the centre of a church spire, a lamp-post, or a weather vane proves to be well situated for indicating the direction of a base line, but its selection possesses this disadvantage, that the theodolite cannot be set up over the point so observed, and when this is the case a station should be staked out in a convenient position upon the base line near the foot of this object, with special reference to the direction of the succeeding base line, so that the angle between the two lines may be measured.

When, as in trigonometrical surveying, it may prove desirable to consider the lofty object first selected for viewing as the main station, the intermediate station is termed a "satellite," and the angle taken from this point requires to be reduced to the centre, *i.e.*, transferred to the proper value for the principal station.

In setting out a line with a theodolite it must be remembered that the effect of the inverting eyepiece is to make the pole or ranging rod appear to require to be shifted to the opposite side of that really needed. Thus, if a pole appears to require moving to the right when viewed through the telescope, it would need the surveyor to give the signal for moving it to the left. This will be clear to the reader upon perusal of the chapter on surveying telescopes in "Field Work and Instruments."

CHAPTER V.

USE OF TRIGONOMETRY.

No survey of any great extent can be conducted without the aid of trigonometrical principles, and although in this country the Ordnance Survey has in a measure done away with the necessity of making surveys of any very great magnitude, still in the Colonies there is yet a field where the principles may be put into practice. During summer months, in order to avoid inflicting injury on farmers by damaging their corn crops in measuring a base line with the chain, the use of trigonometry is of service. When trigonometrical calculations are to be made from the angles, the angles should be read to a single minute. An instrument that will accurately read an angle to a minute will answer very well for a practical surveyor, as angles cannot be laid down nearer to the usual scale of a plan, either by the line of chords or the protractor, but when a theodolite reads to smaller sub-divisions of a degree, it is well to record the same, for although, as stated, the relative bearings cannot be graphically plotted nearer than one minute, the closeness of observation tends to ensure accuracy in a survey. If any two straight lines, such as A C and B C, intersect one another at a point C, and if, from any point in one of these lines, at any distance from C upon the side forming an acute angle with the other line, we draw a perpendicular to meet the other line, a right-angled triangle will be formed. Thus, if from the point B in the line B C, we draw the perpendicular B C to meet the line A C, a right-angled triangle A B C will be formed, in which

— TRIGONOMETRICAL EXPRESSIONS —

DESCRIPTION:—GIVEN ONE SIDE OF ANY RIGHT ANGLED TRIANGLE, TO CALCULATE EITHER OF THE REMAINING SIDES, AFTER ASCERTAINING THE NUMBER OF DEGREES *etc* IN ONE OF THE ACUTE ANGLES OF THE TRIANGLE.———

(Diagram: right-angled triangle with vertices C (angle), B (right angle), A; CB = BASE, AB = PERPENDICULAR, CA = HYPOTHENUSE)

1. $\underline{\text{SIN ACB}} = \frac{\text{PERP}}{\text{HYP}} = \frac{AB}{AC} = \underline{\text{COS CAB}}$
   $AB = AC \times \text{SIN ACB}$

2. $\underline{\text{COS ACB}} = \frac{\text{BASE}}{\text{HYP}} = \frac{CB}{AC} = \underline{\text{SIN CAB}}$
   $CB = AC \times \text{COS ACB}$

3. $\underline{\text{TAN ACB}} = \frac{\text{PERP}}{\text{BASE}} = \frac{AB}{BC} = \underline{\text{COT CAB}}$
   $AB = CB \times \text{TAN ACB}$

4. $\underline{\text{SEC ACB}} = \frac{\text{HYP}}{\text{BASE}} = \frac{AC}{CB} = \underline{\text{COSEC CAB}}$
   $AC = CB \times \text{SEC ACB}$

5. $\underline{\text{COSEC ACB}} = \frac{\text{HYP}}{\text{PERP}} = \frac{AC}{AB} = \underline{\text{SEC CAB}}$
   $AC = AB \times \text{COSEC ACB}$

6. $\underline{\text{COT ACB}} = \frac{\text{BASE}}{\text{PERP}} = \frac{CB}{AB} = \underline{\text{TAN CAB}}$
   $CB = AB \times \text{COT ACB}$

B C will represent the base, A B the perpendicular, and A C the hypothenuse subtending the right angle. (See page 36.)

Mathematical tables have been formed, giving the ratio between any two adjacent sides of a triangle, so that by the aid of these tables, when we know the length of any one side, and the value in degrees and parts of a degree of either of the acute angles A C B or C A B, we can arrive at the length of the adjacent side forming the angle, and then, having arrived at the length of two sides of the triangle, we can easily determine the length of the third side, either by a further use of these tables, or by the application of the 47th proposition of the first book of Euclid.

The trigonometrical values furnished by the accompanying tables for angles, to every degree of the quadrant, are denominated as follows :—(1) the sine of the angle, found by dividing the length of the perpendicular by the length of the hypothenuse; (2) the cosine of the angle, found by dividing the length of the base by the length of the hypothenuse; (3) the tangent of the angle, found by dividing the length of the perpendicular by the length of the base; (4) the secant of the angle, found by dividing the length of the hypothenuse by the length of the base; (5) the cosecant of the angle, found by dividing the length of the hypothenuse by the length of the perpendicular; and (6) the cotangent of the angle, found by dividing the length of the base by the length of the perpendicular.

If in the triangle A B C the hypothenuse A C represents the radius of a circle equal to unity, then the sine of the angle will be expressed by the length of the perpendicular, and the cosine of the angle will be expressed by the length of the base. But if the length of the base be a radius equal to unity, as shown in the diagram, then the length of the perpendicular will express the value of the tangent, and the length of the hypothenuse will express the value of the secant. Hence the names sine, tangent, secant, applied to these expressions, sine signifying a line opposite the angle, and secant a line cutting the circle. The term "sine" is derived from "sinus," signifying a bent surface, or curve, the angle being measured by the sine of the arc.

In the case of the angle C A B, the line C B would become the perpendicular drawn from the point B, at a distance, B A, from the point of intersection, A, to meet the line C A in the point C, and B A would become the base from which the perpendicular would be drawn. Hence the sine, tangent, and secant of the angle A C B would become respectively the cosine, the cotangent, and the cosecant of the complement to the angle A C B; that is, of the angle C A B. Furthermore, it will be observed that in either angle the reciprocal of the sine is equal to the cosecant, the reciprocal of the tangent is equal to the cotangent, and the reciprocal of the secant is equal to the cosine. (See pages 39-41.)

Strictly speaking, angles only determine the relative shape; equiangular triangles may be of very unequal magnitudes, yet perfectly similar, just as a circle may be either that of a watch dial or the orbit of Jupiter. It needs the diameter to be stated to fix its size, so with triangles, we require the length of the sides to be ascertained as well as the angles. But when they are all equilateral, the one having its sides equal to the corresponding sides of the other, they are not only similar and equiangular, but are equal in every respect. The angles, therefore, determine the relative species of the triangle, the sides its absolute size, and consequently that of every other figure, as all are resolvable into triangles. The angles determine direction only, yet the essence of a triangle seems to consist much more in the angles than the sides, for the angles are the true, precise, and determined boundaries thereof. It will also be observed that whatever the size of the triangle, provided the angles remain the same, these ratios will by the principle of similar triangles remain the same; so that when the length of one side is known, the length of the other sides in any particular triangle can be determined.

If a town to be surveyed extends over a considerable area, and affords no position for a long base line, it must unquestionably be surveyed with the aid of trigonometrical observations. A knowledge of plane trigonometry is also

TABLE OF NATURAL COSECANTS

DEC	COSEC	DEC	COSEC	DEC	COSEC
0	INFINITE	31	1·9416	61	1·14335
1	57·2986	32	1·8870	62	1·13257
2	28·6537	33	1·8360	63	1·12232
3	19·1073	34	1·7882	64	1·11260
4	14·3355	35	1·7434	65	1·10337
5	11·4737	36	1·7013	66	1·09463
6	9·5667	37	1·6616	67	1·08636
7	8·2055	38	1·6242	68	1·07853
8	7·1852	39	1·5890	69	1·07114
9	6·3924	40	1·5557	70	1·06417
10	5·7587	41	1·5242	71	1·05762
11	5·2408	42	1·4944	72	1·05146
12	4·8097	43	1·4662	73	1·04569
13	4·4454	44	1·4395	74	1·04029
14	4·1335	45	1·4142	75	1·03527
15	3·8637	46	1·39016	76	1·03061
16	3·6279	47	1·36732	77	1·02630
17	3·4203	48	1·34563	78	1·02234
18	3·2360	49	1·32501	79	1·01871
19	3·0715	50	1·30540	80	1·01542
20	2·9238	51	1·28675	81	1·01246
21	2·7904	52	1·26901	82	1·00982
22	2·6694	53	1·25213	83	1·00750
23	2·5593	54	1·23606	84	1·00550
24	2·4585	55	1·22077	85	1·00381
25	2·3662	56	1·20621	86	1·00244
26	2·2811	57	1·19236	87	1·00137
27	2·2026	58	1·17917	88	1·00060
28	2·1300	59	1·16663	89	1·00015
29	2·0626	60	1·15470	90	1·00000
30	2·0000				

SECANT A = COSEC (90−A)

TABLE OF NATURAL COTANGENTS

DEC	COTAN	DEC	COTAN	DEC	COTAN
0	INFINITE	31	1·6642	61	·55432
1	57·2899	32	1·6003	62	·53170
2	28·6362	33	1·5398	63	·50952
3	19·0811	34	1·4825	64	·48773
4	14·3006	35	1·4281	65	·46630
5	11·4300	36	1·3763	66	·44522
6	9·5143	37	1·3270	67	·42447
7	8·1443	38	1·2799	68	·40402
8	7·1153	39	1·2348	69	·38386
9	6·3137	40	1·1917	70	·36397
10	5·6712	41	1·1503	71	·34432
11	5·1445	42	1·1106	72	·32491
12	4·7046	43	1·0723	73	·30573
13	4·3314	44	1·0355	74	·28674
14	4·0107	45	1·0000	75	·26794
15	3·7320	46	·96568	76	·24932
16	3·4874	47	·93251	77	·23086
17	3·2708	48	·90040	78	·21255
18	3·0776	49	·86928	79	·19438
19	2·9042	50	·83909	80	·17632
20	2·7474	51	·80978	81	·15838
21	2·6050	52	·78128	82	·14054
22	2·4750	53	·75355	83	·12278
23	2·3558	54	·72654	84	·10510
24	2·2460	55	·70020	85	·08748
25	2·1445	56	·67450	86	·06992
26	2·0503	57	·64940	87	·05240
27	1·9626	58	·62486	88	·03492
28	1·8807	59	·60086	89	·01745
29	1·8040	60	·57735	90	·0
30	1·7320				

$$\tan A = \cotan(90 - A)$$

TABLE OF NATURAL SINES

DEC	SINES	DEC	SINES	DEC	SINES
0	·00	31	·51503	61	·87461
1	·01745	32	·52991	62	·88294
2	·03489	33	·54463	63	·89100
3	·05233	34	·55919	64	·89879
4	·06975	35	·57357	65	·90630
5	·08715	36	·58778	66	·91354
6	·10452	37	·60181	67	·92050
7	·12186	38	·61566	68	·92718
8	·13917	39	·62932	69	·93358
9	·15643	40	·64278	70	·93969
10	·17364	41	·65605	71	·94551
11	·19080	42	·66913	72	·95105
12	·20791	43	·68199	73	·95630
13	·22495	44	·69465	74	·96126
14	·24192	45	·70710	75	·96592
15	·25881	46	·71933	76	·97029
16	·27563	47	·73135	77	·97437
17	·29237	48	·74314	78	·97814
18	·30901	49	·75470	79	·98162
19	·32556	50	·76604	80	·98480
20	·34202	51	·77714	81	·98768
21	·35836	52	·78801	82	·99026
22	·37460	53	·79863	83	·99254
23	·39073	54	·80901	84	·99452
24	·40673	55	·81915	85	·99619
25	·42261	56	·82903	86	·99756
26	·43837	57	·83867	87	·99862
27	·45399	58	·84804	88	·99939
28	·46947	59	·85716	89	·99984
29	·48480	60	·86602	90	1·00000
30	·50000				

COSINE A = SINE (90−A)

useful for calculating distances across valleys or rivers over which a chain cannot be stretched, also for arriving at the co-ordinate dimensions to be employed in plotting a traverse survey (page 85). The accuracy of the work depends upon the accuracy with which observations are made with the use of an optical and angular instrument, nature providing in its

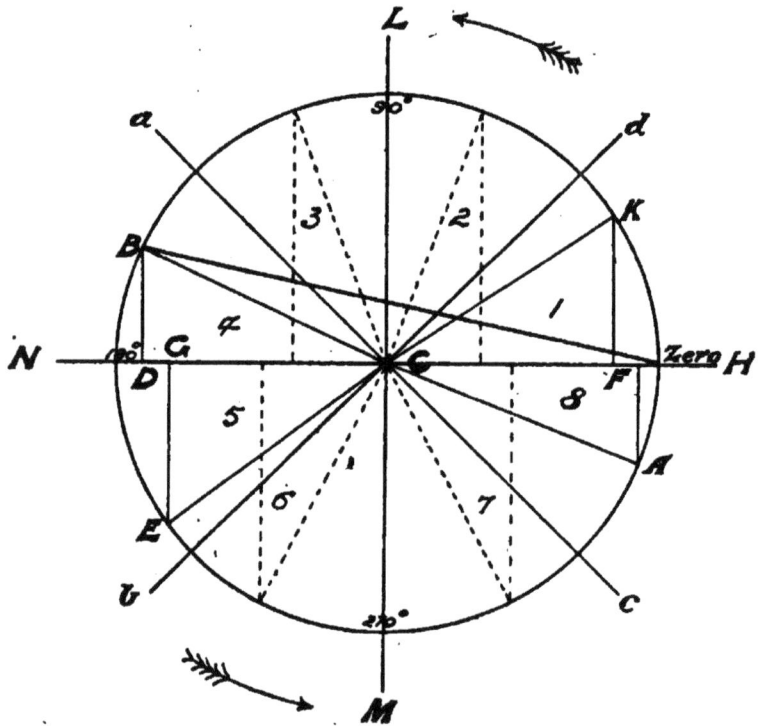

rays of light a practical infinity of straight lines which stretch in every unobstructed direction, and which may be regarded as the sides of a series of triangles which form the optical base lines of a survey. A knowledge of plane trigonometry is particularly serviceable in tunnel work and in mining surveying, and is employed in measurements for uniting any detached underground survey to a surface survey, for the purpose of ascertaining the extent of workings in regard to adjoining property; also to determine the height and distance of objects relatively to one another.

In ordinary practice, where minutes and seconds of an angle are recorded as well as degrees, a table of logarithms is indispensable for the solution of trigonometrical problems. The signs of the trigonometrical expressions are also necessary to be understood. Suppose N C H and L C M to be two lines at right angles to one another, and fixed in position, and that the radius C H traverses the circle in the contrary direction to the hands of a watch, as indicated by the arrows; in doing so it passes through the sectors of the circle marked 1, 2, 3, 4, 5, 6, 7, 8 in the figure (page 42). Assuming H to be a point upon the circumference of the circle at zero on a scale of degrees, if radius $C H = a$, and $C F = x$ and $F H = b$, then $x = a - b$. Now, so long as $b < a$, x is $+$, and F lies to right of C; but if $b > a$, x is $-$ and lies to the left of C. Hence any line measured along N H or parallel to it, is said to be a positive line on the right of C, or a negative line on the left of C, that is along C N or parallel to it, and the symbol $+$ or $-$ represents the direction. The radius lying in the direction of neither the vertical nor horizontal direction cannot change its sign, and is always reckoned positive. Hence the sign of the trigonometrical expressions will be as follows:—

In 1, 2, 3, 4 the sine is $+$
In 5, 6, 7, 8 the sine is $-$
In 1, 2, 7, 8 the cosine is $+$
In 3, 4, 5, 6 the cosine is $-$

and these signs determine the signs of the remaining trigonometrical ratios of tangent and cotangent, secant and cosecant. In dealing with an angle a, the expression $\cos a$, &c., &c., will be independent of the magnitude of the revolving line and depend only on the absolute inclination of the two lines.

Since $\cos a$ is never greater than unity, vers a is always $+$ and its greatest value is when a becomes $180°$ when $\cos a = -1$ and vers $a = 2$.

Dividing the circle into eight parts as shown in the figure we have

	The greater.	The less.	Angle.
1.	cos α +	sin α +	0° to 45°
2.	sin α +	cos α +	45° to 90°
3.	sin α +	cos α −	90° to 135°
4.	cos α −	sin α +	135° to 180°
5.	cos α −	sin α −	180° to 225°
6.	sin α −	cos α −	225° to 270°
7.	sin α −	cos α +	270° to 315°
8.	cos α +	sin α −	315° to 360°

It is useful to remember also that the
  sin of an angle + sin of another angle
  = 2 sin semi. sum (cos semi. diff.)
  cos of an angle + cos of another angle
  = 2 cos semi. sum (cos semi. diff.)
  sin of an angle − sin of another angle
  = 2 sin semi. diff. (cos semi. sum)
  cos of an angle − cos of another angle
  = 2 sin semi. sum (sin semi. diff.)
Also that
  2 sin of larger angle (cos of smaller angle)
  = sin of sum + sin of diff.
  2 sin of smaller angle (cos of larger angle)
  = sin of sum − sin of diff.
  2 cos of one angle (cos of another angle)
  = cos of sum + cos of diff.
  2 sin of one angle (sin of another angle)
  = cos of diff. − cos of sum
  If $A < 45°$ cos A > sin A
  If $A > 45°$ and $< 90°$ sin A > cos A.

The student desirous of working out one or two practical examples in the application of plane trigonometry to surveying, will find the following questions useful to test his proficiency.

(1.) Having measured B A up a slope, and taken the

angles D B A, B A D, C B A, C A B, show how to calculate the lengths of B C, B D, and C D.

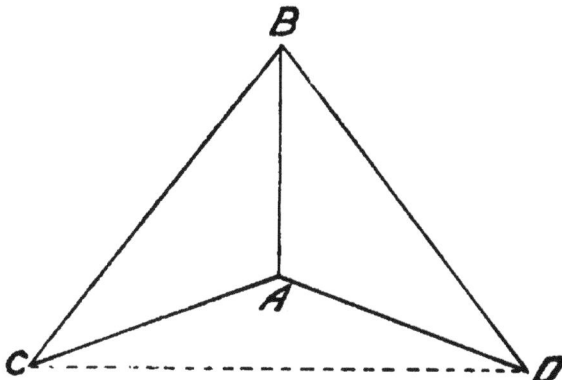

(2.) Two men are surveying when each is at a distance of 300 yards from a flagstaff; the one finds the angle subtended by the position of his companion and the staff to be 32° 45'. Find how far they are apart.

(3.) In walking towards a flagstaff, a surveyor found the angle of elevation of its top to be 2° 19' 13" at one milestone, and after proceeding to the next milestone, the angle of elevation was 3° 28' 49". How much farther should he have to walk before he reached it, assuming they were all on the same level?

(4.) Wishing to know the height of a certain house standing on the summit of a hill of uniform slope, a surveyor descended the hill 40 feet and then found that the height subtended an angle of 34° 18' 19", but upon descending a farther distance of 20 feet, he found this angle had become 19° 14' 32". Find the height of the house.

The student who can work out these problems, and also is able to follow the trigonometry involved in the formulæ for the setting out of curves, given on pages 203-225, will be able to deal with any problem in plane trigonometry likely to come before him as a surveyor.

CHAPTER VI.

FIELD-BOOK.

THE field-book of a survey should contain upon its first page the surveyor's name conspicuously written, to provide for its chance of recovery, if lost, an index to the base lines, not necessarily drawn to scale, but sketched so as to indicate the relative position of the main base lines, and the direction in which they are chained. Thus the index sketch of the base lines used in plotting a survey made from chain measurements only is represented in the accompanying diagram (pages 47-49). The base lines are numbered, and an arrow upon each line shows the direction in which they were measured, in the field or upon the ground.

Pencil entries are preferable to the use of ink in the field, because the ink will run in wet weather. No note should be left to memory, but should be written or booked at once. The surveyor should always carry a piece of india-rubber in his pocket. Few can sketch correctly, as well as clearly, without the use of india-rubber. Such corrections would mean erasing if pen and ink were employed. Measurements should be written small but clear. There is no occasion to chain a base line twice in order to prove its accuracy. The tie-line is the best proof of correct measurement, and is founded upon a proposition of Euclid (book 1., prop. 7), that upon the same line, and upon the same side of it, there cannot be two different triangles, having their sides terminating at the same extremity of the base, equal to one another. Thus the line marked 7 in the diagram, proceeding from a point in the line A B to the intersection of lines marked 3 and 6, forms a tie to the triangle formed by the lines marked 1, 3, and 6 respectively.

Scarcely any two surveyors set down their field notes exactly in the same manner. Usually, however, the main terminal stations are lettered in the field-book, but not the

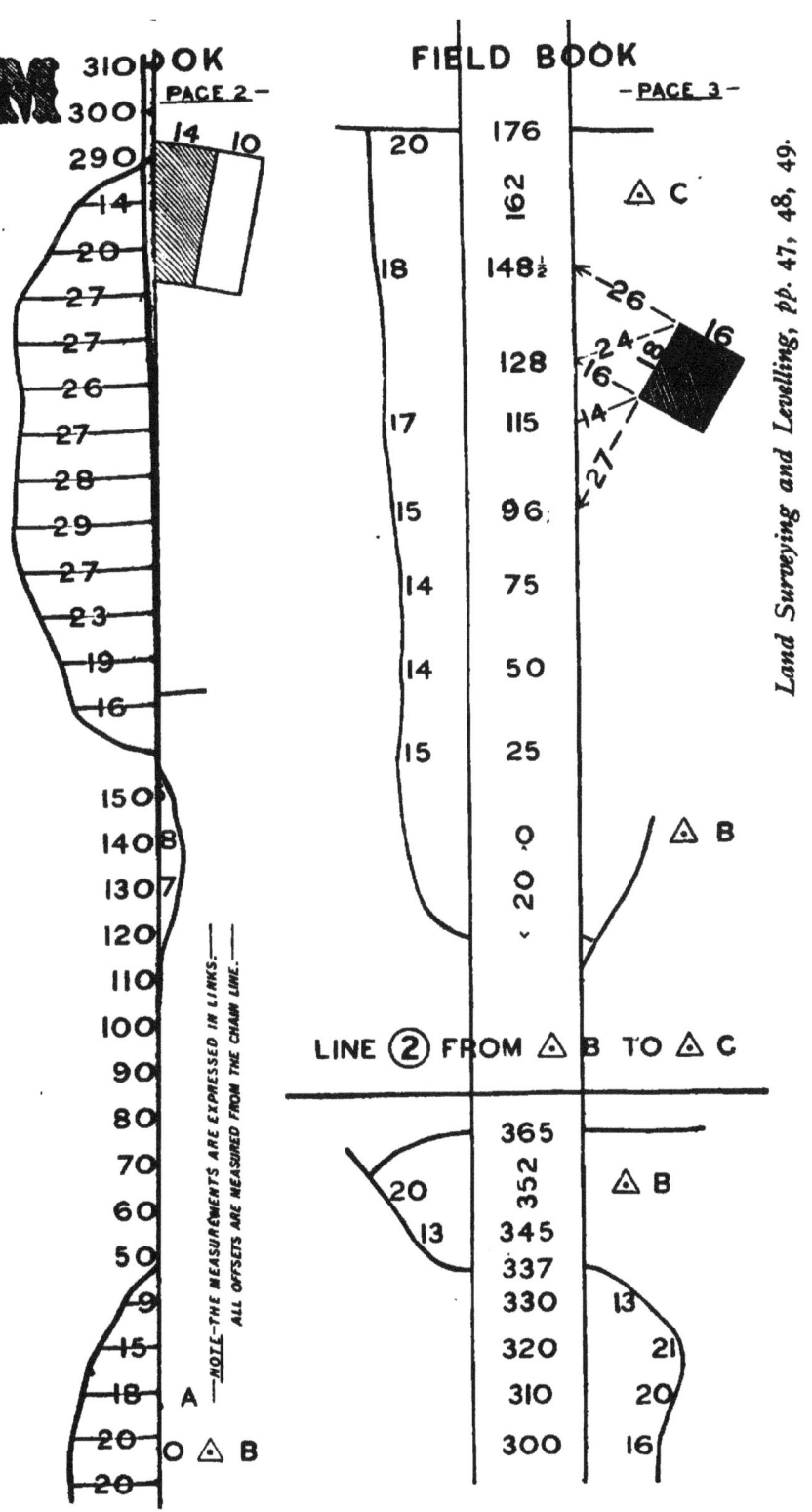

intermediate stations, otherwise in an extensive survey the alphabet would soon be exhausted. A peg is driven into the ground at each station, and the chainage is always started from the peg, which denotes the station. Thus, in line 2 from station B to station C, the zero of the chain is placed at B when chaining the line B C. The position of the fence, twenty links back, can be most accurately determined by measuring this distance in the direction of the line B C, but this is done with the tape or with the chain, before measuring line 2. The total length of line 1 at station B is 352 links from station A. The details of keeping a field-book vary very much in the practice of individual surveyors, but in the Author's plan the total length at a station point is written in the field-book sideways, as shown, to distinguish it from the other chain measurements. The line is then continued on, to meet the fence at 365 links. In line 2 the measurement is continued on in a similar manner to meet the fence at 176 links.

The surveyor should endeavour to have as few pegs as possible near each other in filling in the details of a survey, a station from which several lines radiate being more correctly determined than where several single lines commence or terminate close to one another upon a base line. The exact distance of an intermediate station from the commencement of the base line may be best distinguished by enclosing the length entered in the field-book in an oval or oblong sketched in pencil, and placing a small circle with a dot in the centre upon the side, right or left, in which the branch line is to be chained. This is shown in the case of line 7, which is measured from 170, in the line 1, to the right hand of A B. It is not necessary to note the distance of all the poles employed for giving the direction of a base line. Only the chainage of those used for subsidiary triangles and side base lines need to be recorded.

The most usual form of field-book is that indicated in our diagram as pages 1, 2, and 3, with two central lines (see pages 47-49), generally ruled in blue, upon each page of a field-book. As a rule it is not necessary to page a field-book. The leaves of the book should be regarded as forming a continuous roll, divided into pages solely for convenience of binding, and to be suitable for carrying in the pocket. The surveyor commences his entries at the bottom of a page, and works upwards as he proceeds along the base line, a method which a little reflection shows to be quite natural, and preferable to all others, as it places the survey-book in the same relative position as the base lines with respect to the surveyor, who advances towards the distant station. Thus the bottom of page 3 follows immediately the top of page 2 of the typical field-book which we have furnished. (See pages 47-49.)

To render the central column of the field-book intelligible, we will assume that a draughtsman could have no difficulty in plotting the irregular fence line which crosses the line marked M M, if perpendicular measurements are given, as shown, at distances say 10 links apart along the line M M. If this straight line, M M, be supposed to be widened out into a column, N N, sufficiently broad to write in the measurements taken along the line M M, and the offsets varying from 7 links to the right of 60, up to 29 links to the left of 210, be repeated, we have practically in the form N N a field-book, as indicated by page 2 in the diagram.

In keeping the field-book, therefore, it should always be remembered that the central column is virtually a representation of a chain line, and that the field-book should otherwise be as much as possible a fac-simile of the ground itself, with every post, hedge, house, pond, &c., placed on the pages of the book with regard to the central column, as they exist with regard to the chain line on the ground. It will be observed that the vertical lines for the offsets, measured from the line M M, or the column N N, are omitted in a field-book, and the distances only are figured, against the line indicating the fence, it being assumed that they are at right angles to the base line, unless otherwise sketched. As the central column is simply provided for the convenience of clearly recording the chain measurements taken along the base lines, and therefore practically indicates a single line, the fence line at 50, 160, and 290 links respectively, where it crosses the base line A B, should *not* be drawn across this column, but should be sketched, as shown, upon each side of the column, and the measurement at the intersection upon the chain stated in links by figures within the column, not in chains and links; thus the entry 956 would read, when plotting, 9 chains and 56 links, but the decimal point is never inserted in the field-book. When the 100 feet chain is used, the offsets are best booked in feet and fractions of a foot, not in feet and inches, thus: 10 feet, $10\frac{1}{4}$ feet, $10\frac{1}{2}$ feet, $10\frac{3}{4}$ feet, 11 feet; since an entry 11' 0" might confuse with 1' 10" in plotting, and similarly in the case of other entries. Inattention in this particular causes much confusion in the relative position of offsets. The exercise of judgment is needed when taking the offsets, so as to select their correct positions, in order to avoid taking an unnecessary number of offsets.

As the chaining goes on, the surveyors mark and note the distances from the commencement, in each case, at which intermediate stations upon the main line would be suitable in the survey, also as in the case of A C and B D, the intersection of other base lines and the crossing of all fences and boundaries. Small triangles and other offsets not rectangular may best be recorded by sketching a diagram of the lines so measured and writing their lengths along them. Tie lines, instead of being measured exclusively for such purposes, should, where possible, be arranged so as to take up such features in the ground as it would be necessary to represent upon the plan. At the same time, whatever distance will in any way help to confirm the accuracy of base lines should be measured by the surveyor. Incorrect results, however slight, are a cause of much anxiety and perplexity. In case one tie line happens to be incorrectly measured or described, the surveyor should possess a proof of his work in the record of the length of other tie lines.

A small triangle with a dot in the centre is the mark adopted by the English Ordnance Surveyors for distinguishing their trigonometrical stations, and is recommended by the author for all main stations as shown in the field-book given in the diagram, pages 47-49. Lettering △ for reference is better than saying (as some books teach) from "end of line No....'; thus in the case of line 3 it is run from the end of line 2 and also from the end of line 5. The latter description would therefore be ambiguous.

CHAPTER VII.

POSITION OF BASE LINES.

In open country the relative position of survey lines can be best fixed by chain measurements. In laying out base lines it must be remembered that the triangle is the only geometrical figure which is incapable of change of form without altering the lengths of the sides, and that when the position of the sides is proved by tie lines, it shows that the lengths of the base lines have been correctly measured. We shall allude to this more in detail when describing the means adopted for plotting a survey in a future chapter. A four-sided figure, like the frame of a slate with the slate out, can assume a variety of shapes without altering the lengths of the sides, but as soon as the corners are diagonally tied together, triangles are formed, and the position of the sides is fixed. In making a survey it is better to work from the whole to a part, than from a part to the whole, not adding field to field and acre to acre, but to embrace the whole survey in one grand system of triangulation. The longest base line should if possible extend throughout the whole length of the survey, and the exterior boundary be measured by offsets from base lines, which should be connected by triangulation to the longest line. The interior details can then be obtained by the formation of subsidiary triangles, and the measurements taken are recorded in a manner described in the previous chapter under the head of Field Book.

Experience alone can suggest the best position for the base lines, the angles between them being neither too obtuse nor too acute, but as equilateral as possible, as otherwise a very trivial error in the plotting of any one of the sides will materially alter the whole figure through the intersection being not well determined, and consequently

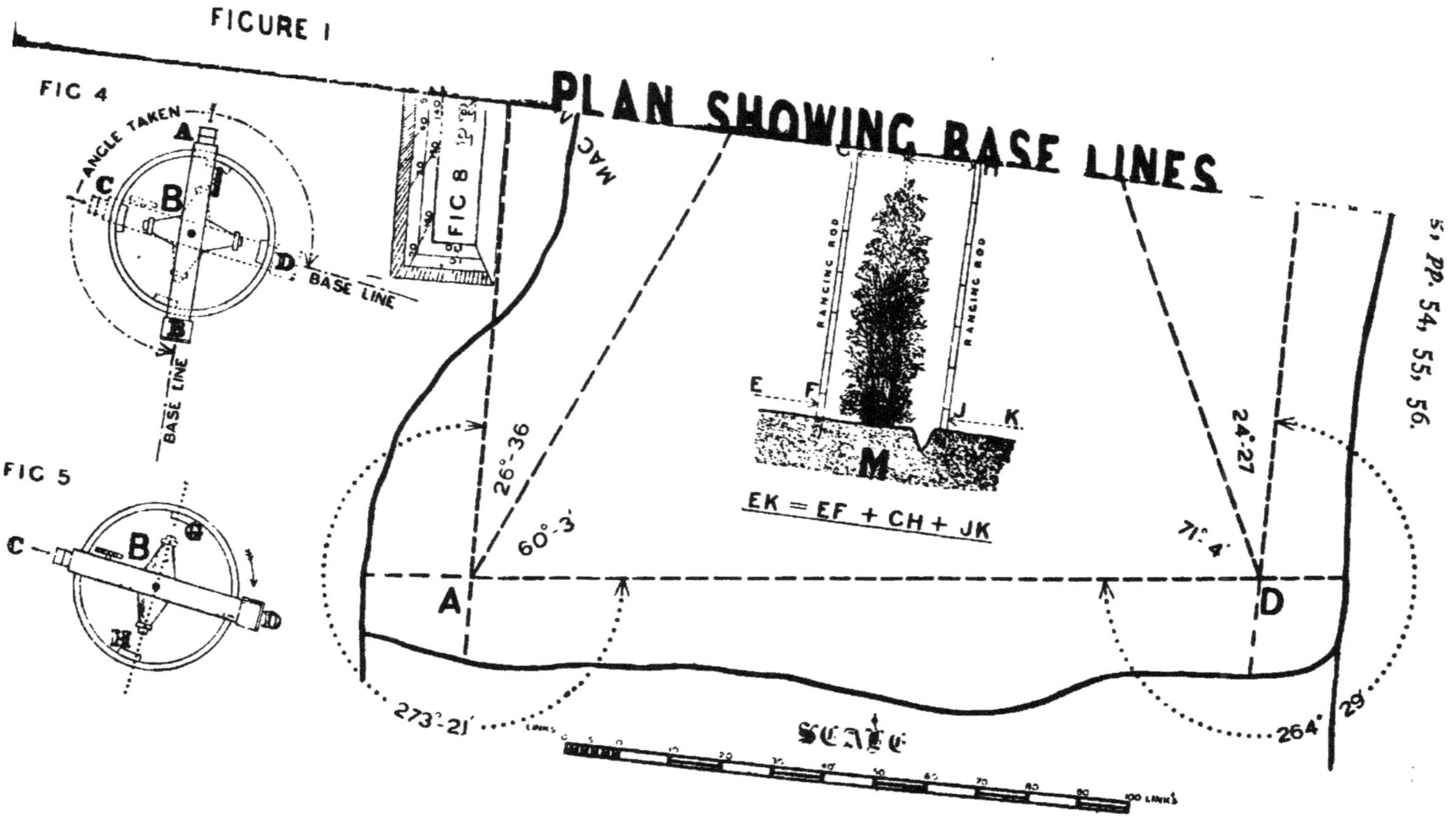

FIGURE 1

PLAN SHOWING BASE LINES

5, pp. 54, 55, 56.

82°
264°
24°·27'
71°·4'
D
$EK = EF + CH + JK$
M
E F   J K
RANGING ROD
60°·3'
26°·36'
MAC
A
273·21'
Scale
00 LINKS

FIG 4
BASE LINE
A D BASE LINE
C B
ANGLE TAKEN

FIG 5
C B A

the area within it will be altered. Although no other fixed suggestion can be made, yet triangulation in surveying is by no means arbitrary; the merit lies in good construction, which may sometimes involve trespassing in a neighbouring field (see line A B, pages 54-56). In the case of a trespass, a favourable impression is created by the observance of common civility only, while offering a full explanation of matters, and at the same time seizing this opportunity for making inquiry upon any points of ownership required.

The surveyor will require the assistance of at least three men, two at the chain, and one to hold the tape and to look after the poles or ranging-rods. It is a great mistake to work short-handed, as it leads to a waste of the surveyor's time in running about. He should so train his men at the outset that in order to keep his field-book clean, it may not be necessary for him to handle the chain while at work, but he must observe that the leader of the chain always holds the arrow denoting the measurement upon the ground perpendicularly, and in a right line with the object of direction, otherwise the line will not measure the true total length between the stations. The precautions mentioned in Chapter III with reference to the chain must be likewise observed. Rather elevated points in open ground (see p. 10) are to be preferred in selecting station points and positions not likely to be disturbed by the plough or by cattle, so as to be available for future reference until accuracy be proved by plotting. In fig. 1, pages 54-56, the fences are shown by full lines, and the base lines are indicated by dotted lines. When base lines cross a wall fence, the crossing points may be notched with a chisel. The lines from which offsets are to be measured, should pass near to or intersect the fences and boundaries, but it would be a mistake to spoil a well conditioned triangle for the sake of running one of its sides near a fence, and hence subsidiary triangles are added to the main triangulation, in order to take offsets from the chainage of the sides of the subsidiary triangles, when the sides of the main triangle are too far off for offsets of reasonable length. Straight fences or walls may be taken up by a process termed "sighting," which consists in taking offsets for plotting, to each end of their line of direction.

In this case the position of the extremities must be very accurately determined by more than one measurement from the chain, as indicated by the triangulation off the base line shown at the intersection of the fences marked J in fig. 8 (pages 17 to 19), and in the measurements indicated by the lines run off the base lines A B and B C, to the outbuildings shown upon the accompanying plan (pages 54 to 56). Triangulation is here adopted in order to determine the exact position with greater accuracy than is attainable by offsets approximately rectangular. Wherever the slightest bend appears to exist in a boundary, an offset should be taken to it from the chain, in order to show the actual outline upon the plan, as indicated at the points marked E, F, G, and H, in fig. 8 (pages 17-19). The offsets are measured from either side of each chain's length, simultaneously with the chaining of the base line, while the chain lies between two arrows upon the base line. So long as the total length of the chainage of a line is correct, it does not matter to the precise length of a link where the offset is taken, and therefore the chain is allowed to lie slack, but straight, without being kept tight, as it is impossible with the use of an ordinary scale to plot to single links. A bundle of, say, stout laths, as used by plasterers, divided into half lengths and pointed at one end becomes very serviceable for ranging, and being light are easily carried about until required. In chaining the line from station C to station A (see pp. 54-56), if a lath is left upon the line at 150 links, and another lath placed at the intersection of the hedge at 190 links, it will be unnecessary to re-set out this line when measuring the line D B, in order to determine, if desired, the intersection of the two diagonal lines, as the tape stretched from one lath to the other would cross the chain upon the line D B at 30 links from the first lath in the line C A, showing the intersection of the two tie-lines to be at 180 links from C. On no account should any important point on the survey be trusted to a single offset. In continuing the measurement of a base line through a hedge, the chain should, if possible, be pulled through, or if the hedge be too thick, as supposed to be the case at M in the line D B, the process illustrated by fig. 6 (pages 54-56) is adopted, care being taken o place the two poles accurately in line and vertical, chaining orward from E to F, taping the distance between the poles over the hedge, and then adding the measurement to E F, and placing the proper link at G to indicate the amount so determined before continuing the measurement to H.

In the case of a town survey, never attempt in crowded thoroughfares to survey both sides of a street off one base line. The chain will be least disturbed by traffic, if lines are run as near as possible along either kerbstone, or close to the fronts or areas of the houses. (See page 62.)

The magnetic bearing of one of the base lines by which the relative position of the north of the survey is arrived at, may be taken with a theodolite which contains a compass, or with a prismatic compass. If the prismatic compass held in the line A B reads 34°, as shown upon the plan, the bearing can be plotted with a protractor, as shown in fig. 8 (pages 54-56), allowance being made for the deviation of the true north from the magnetic north. If the prismatic compass be held upon the line C A, and viewed in the direction from C to A it will read 242°. A reference to the article "Magnetic Elements" in "Whitaker's Almanack" for the current year will give the variation from the true north.

It is often advisable to lay down the base lines of a survey roughly to a small scale upon a sheet of drawing paper, in order to determine therefrom the exact position for the base lines either to be set out or for purposes of plotting. It frequently happens that a sketch has to be made when no instruments are obtainable, but by means of a single sheet of paper, a pencil, and a straight-edged ruler or piece of wood, such as could always be procured, very valuable work can be produced. In sketching without instruments care is more than ever essential, and it must be always remembered that the principle on which surveying is based is invariably the same whether instruments are employed or not.

Before measuring a base line it must be set out. Upon an extensive survey, the theodolite may be employed in setting out long base lines, provided it is in thorough adjustment, so that in ranging out the required direction the vertical circle of the instrument moves up and down in a direct vertical plane, and the cross wires clearly define every point fixed in a straight line with the centre of the telescope. The surveyor should have a small flag upon a ranging rod to direct his assistant when he is at too great a distance from him to hear or to see a motion of the hand. Referring to our diagram (pages 54-56), let us suppose we require the angle between the base lines A B and B C, in fig. 1. The instrument when removed from the box is in an unclamped condition, and is set up over the peg at B, before in any way clamping the plates. The vernier of the horizontal plate is then set to zero at F in fig. 2, and clamped. The upper plate being thus clamped to the lower plate, the reading at 360° is finally adjusted by the tangent screw with the aid of the magnifying eye-piece attached to the horizontal plate. Both plates thus clamped are revolved upon the vertical axis by turning the telescope with the hand until it comes as nearly as possible in the direction of the line B A, and the line of collimation in the telescope is set to intersect the centre of a ranging-rod or peg, which has been previously placed exactly upon the line B A. The collar which fixes the lower plate or limb of the instrument is then clamped, and the lower milled-headed tangent screw connected with this collar clamp is slowly turned by the hand until the distant point of observation upon the base line is accurately bisected. Having done this, leave the lower plate clamped, and unclamp the vernier plate. By this means, freedom will be given to the stage carrying the telescope to traverse the required angle. The verniers move simultaneously with the telescope, and the angle is measured by the amount of arc described, after the telescope is set exactly in the direction of the line B C, and the line of collimation bisects the distant point to be observed. In proceeding from station A to station C in fig. 1, the angle between the line A B and B C is booked at 278° 11′, whereas if we were proceeding in the opposite direction from station C towards station A, the angle at B would be booked at 81° 49′. For convenience of reading the verniers the zero point is placed at the side of the stage carrying the telescope, so that while the angle required to be taken is situated as shown in fig. 4, the actual angle traversed by the verniers is shown in fig. 3. Having clamped the instrument, the surveyor reads the degrees and sub-divisions marked on the primary circle up to the arrow upon the vernier scale, and next proceeds to pass the eye along the vernier scale of the instrument towards the left, and then reads the additional number of minutes off the vernier or travelling scale without any reference to the primary circle, except for the coincidence of one of the vernier scale divisions with any division upon the primary circle. In all angular measurement the prevention of error is dependent upon the power and quality of the instrument employed, and is independent of the space over which the work extends. It is in the execution of a survey of a large estate that great judgment is required as to the actual angles to take, and as stated elsewhere (page 90) the position of the meridian line should be made but a secondary consideration, and in every case to depend upon the shape and size of the plan., Where the two can be united, the better, as will be found to be the case in Ordnance maps (pages 132, 133).

The dimensions of villages, towns, and cities may be obtained by the chain only, if the streets are wide enough to admit of the angle at the meeting or intersection of the roads being taken by tie lines. When the main lines are to be so connected, and especially if they are of some length, it will be necessary to run such tie lines as far away as possible from the point of intersection, for a small error in laying down the plan with short tie lines would cause the main lines to deviate considerably from their true position when prolonged. To secure accuracy in plotting it is advisable in such cases to measure the tie lines to a quarter of a link, and then to multiply both them and the distances from the angular points by 2, 3, 4, or any larger number as circumstances may require, and to use such products in laying down the base lines. (See figs. 9-11, pages 28-30.)

Where a survey is required for the purpose of setting out building lots or for executing any new work which may extend over a considerable period, it is advisable that the surveyor should leave upon the ground good permanent

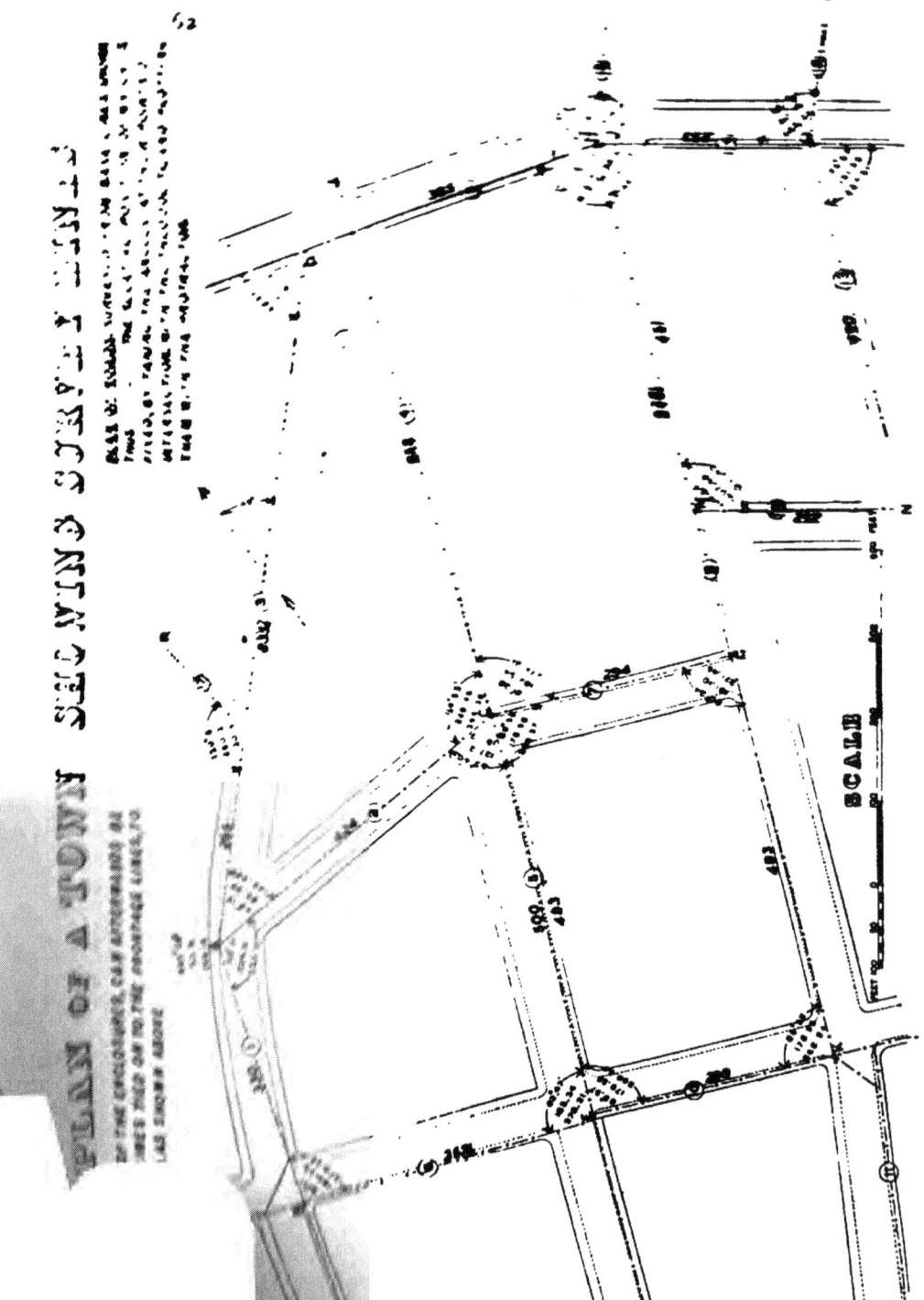

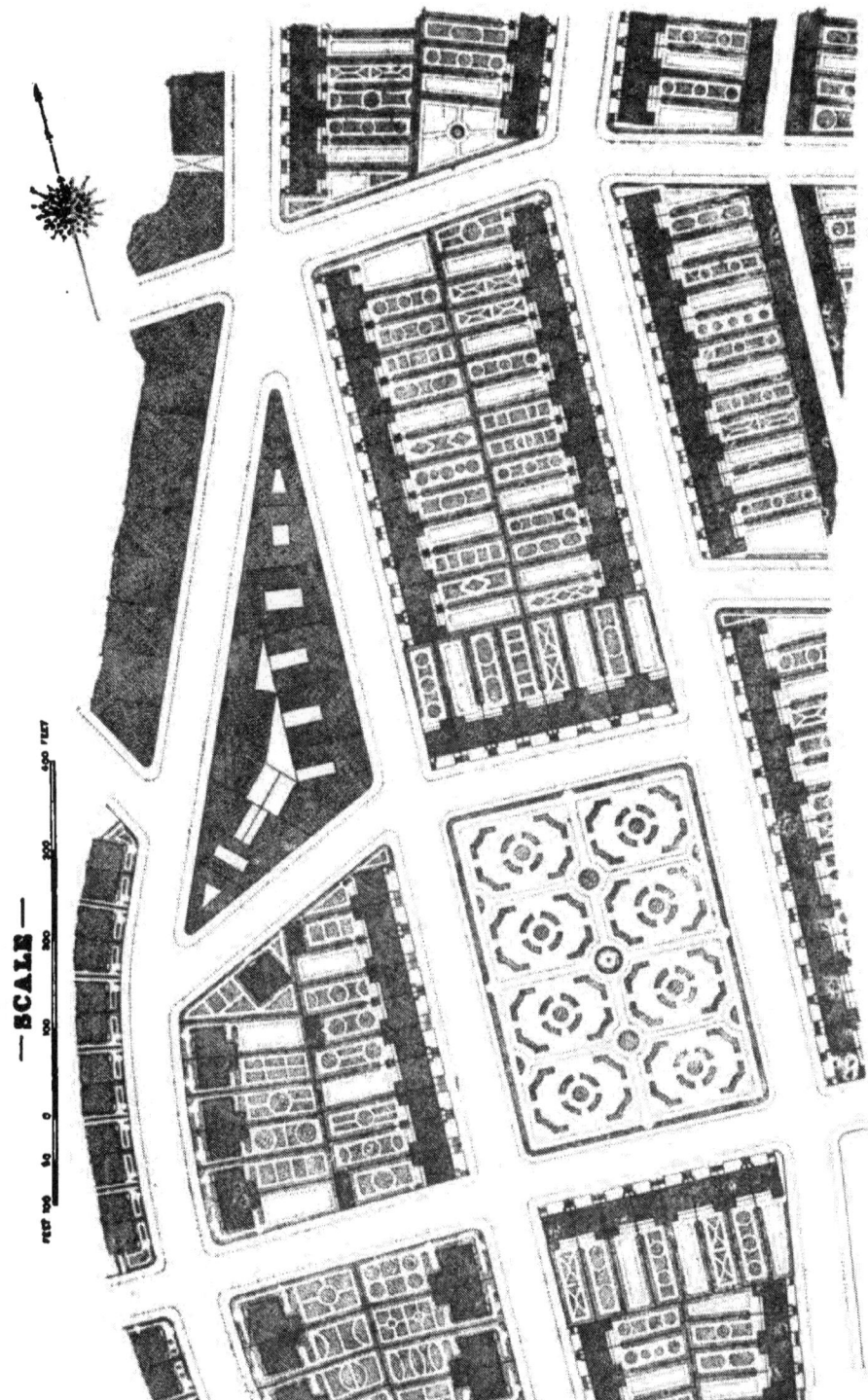

PART PLAN OF A TOWN
— SCALE —

PART PLAN OF A TOWN SHOWING SURVEY LINES

NOTE—THE DETAIL OF THE ENCLOSURES, CAN AFTERWARDS BE SURVEYED IN, FROM LINES TIED ON TO THE FRONTAGE LINES, TO COMPLETE THE PLAN, AS SHOWN ABOVE

PLAN OF ROADS, SURVEYED FROM BASE LINES SHOWN THUS ———— THE RELATIVE POSITION OF WHICH, IS FIXED, BY TAKING THE ANGLES AT THEIR POINTS OF INTERSECTION, WITH THE THEODOLITE, AND PLOTTING THEM WITH THE PROTRACTOR. ————

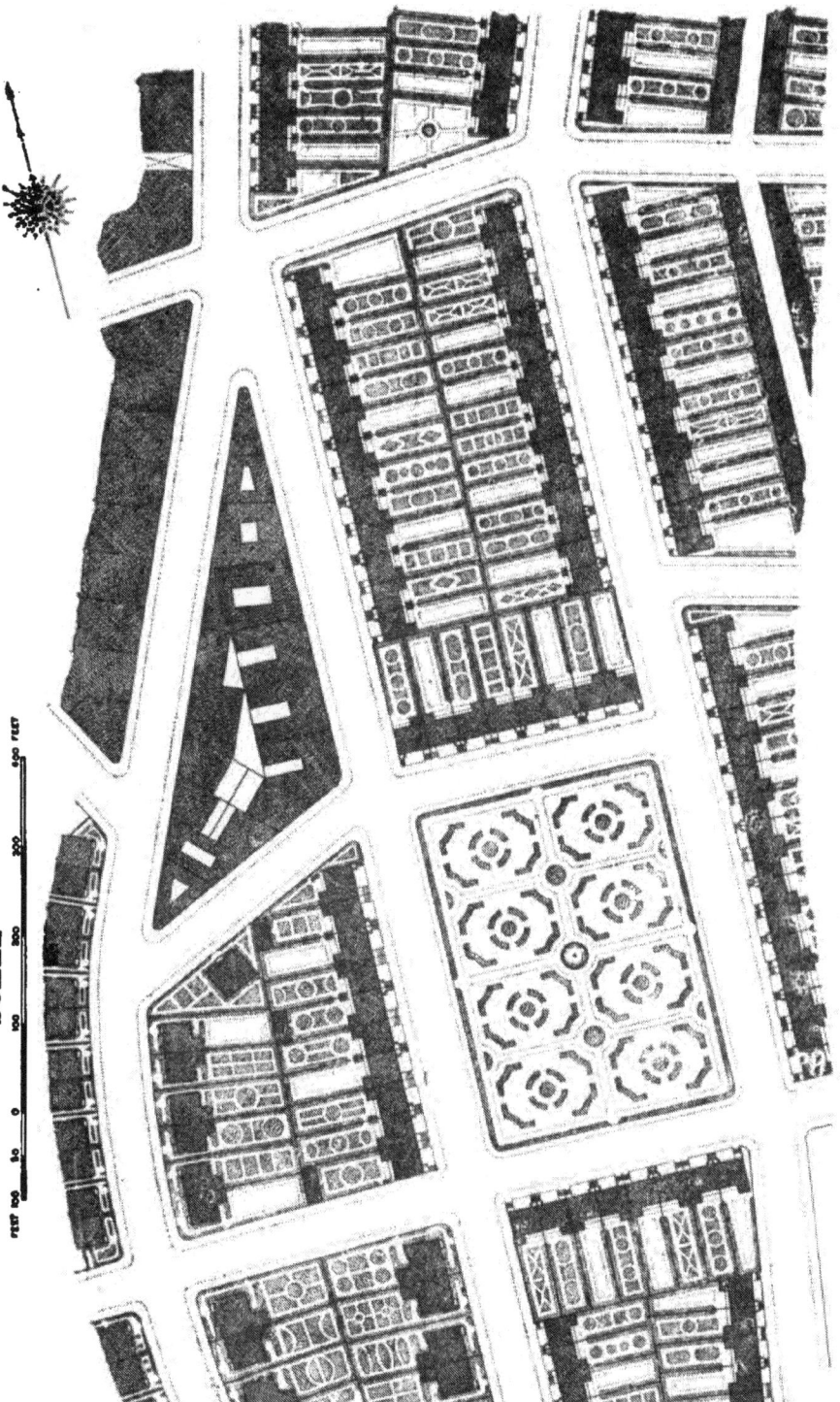

PART PLAN OF A TOWN
— SCALE —

and well-defined stations, so that all his principal lines of construction may be re-set out at any future date in exactly the same position as when the survey was originally made, as it is necessary in order to set out allotments correctly upon the ground always to measure from the lines of the survey, and not from the fences. To record the line for such purpose, it is well to drive nails into the tops of the stakes or pegs and to strain a line of fine cord or wire from nail to nail, then with a sharp spade to nick out the line required as in setting out the centre line of a railway.

When a long line of country has to be surveyed, a separate portion is usually allotted to different surveyors, each part having to be connected at its junction. Accuracy in connecting the base lines at such junctions will be aided by continuing them so as to overlap each other's work, and fixing them by intersecting triangles. It is always better to have a line or angle too much than to find difficulty in plotting by having omitted what may have appeared in the field to be a very trifling line or angle. The element of uncertainty should be entirely obliterated, and it is better to sketch too much than too little.

Referring to the plan (fig. 1, pages 54-56) showing base lines, the surveyor may make a short prolongation of a long fixed line as shown at the angles A, B, C, and D in the diagram, but he should never make a long prolongation of a short line. When the produced portion is over a chain in length, the end of the line so produced should be connected by a tie line. (See fig. 11, pp. 28-30.) In the insertion of a detail plan of a building the dimensions are taken and entered in feet and inches upon an enlarged plan, made upon a separate page and properly referenced in the ordinary field entries. In order to furnish a correct plan, the thickness of walls, the length and breadth of rooms and passages, the width of doors and windows, the projection of fireplaces and other necessary dimensions need to be booked. If the premises are to be sold, every convenience should be noted and marked upon the plan in order to promote the sale, and it will be found very advantageous to have plans of the cellars and of the upper stories, and in some cases a sketch of the elevation, as well as a photograph thereof.

POSITION OF BASE LINES.

Where a system of chain triangulation cannot be adopted, the angles between the base lines may be taken with a theodolite in the following manner :—In surveying a town, where possible, commence the survey at the meeting of three or more of the principal streets through which the longest prospects can be obtained for the purpose of laying out base lines. The instrument is set up level with its vertical axis exactly over the intersection of two base lines. This is indicated by the plumb-bob hanging exactly over the centre of the station peg, the surveyor exercising the precautions as to the exact point, which were stated in Chapter II., figure 8 (pages 6, 7). For this purpose some surveyors use wire, others think a chain preferable on side-long ground, while the majority prefer whip cord to hold the plumb-bob. Upon a well-paved road the best way to mark the station points is to run the base lines so that they terminate, where possible, in the joints of the kerbstone, into which a 4 inch nail or stout spike can be easily driven (see fig. 10 pages 99-101), and which can be readily found when required for future reference, by noting its distance along the kerb from the nearest lamp-post, or by triangulation from other fixed points, as otherwise if the streets in which stations are necessary should be paved or pitched it would be impossible satisfactorily to mark the precise spot. The preceding diagram illustrates a part plan of a town where this method was adopted. (See page 63.) The roads were first surveyed from the base lines as shown. (See page 62.) The angles at the points of intersection are taken, the subdivisions being read by means of a vernier scale, and the lengths of the base lines accurately chained. The detail of the enclosures can then be afterwards surveyed from base lines fixed in position by being tied on to the frontage lines to complete the plan. (See page 63.)

The theodolite is the only perfect angular instrument with which horizontal bearings can be taken in an undulating country. The accuracy of the horizontal angles which are measured is proved by a process of repetition, which consists of respectively doubling and trebling, mechanically, the angle first read, and in order to avoid error of direction in plotting, it is advisable to adopt a uniformity of method, by always recording the amount of arc traversed by the

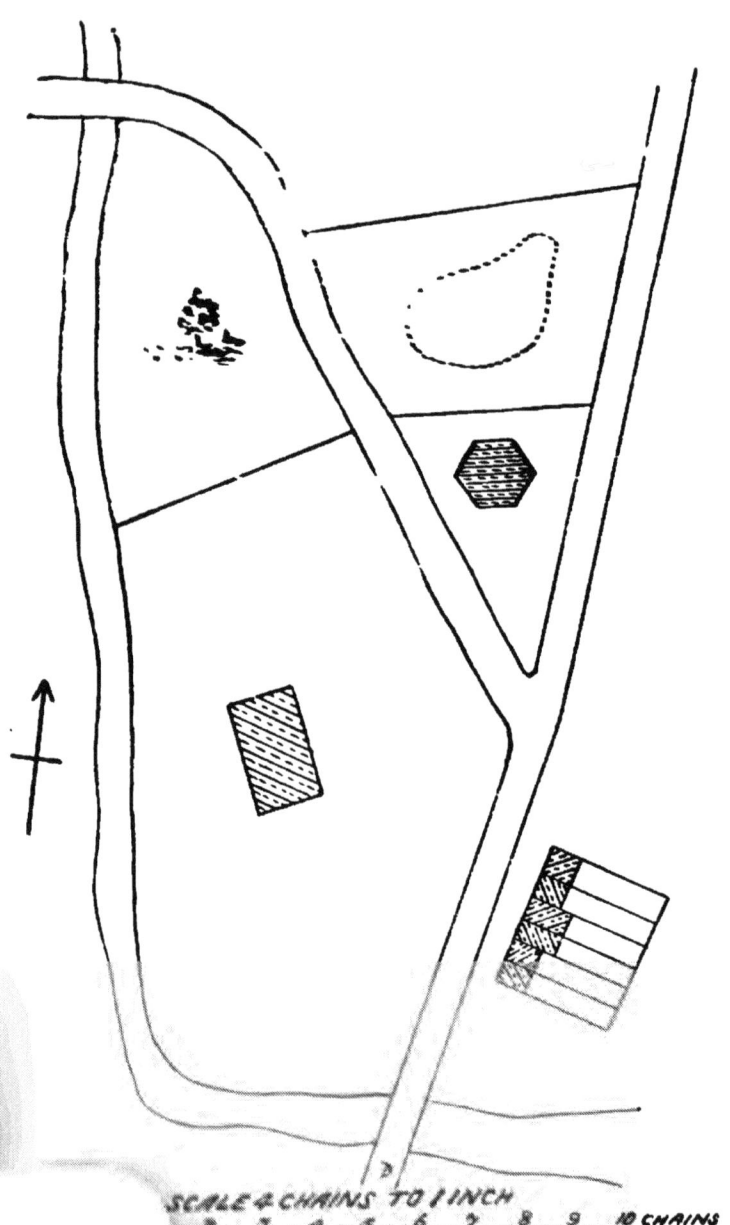

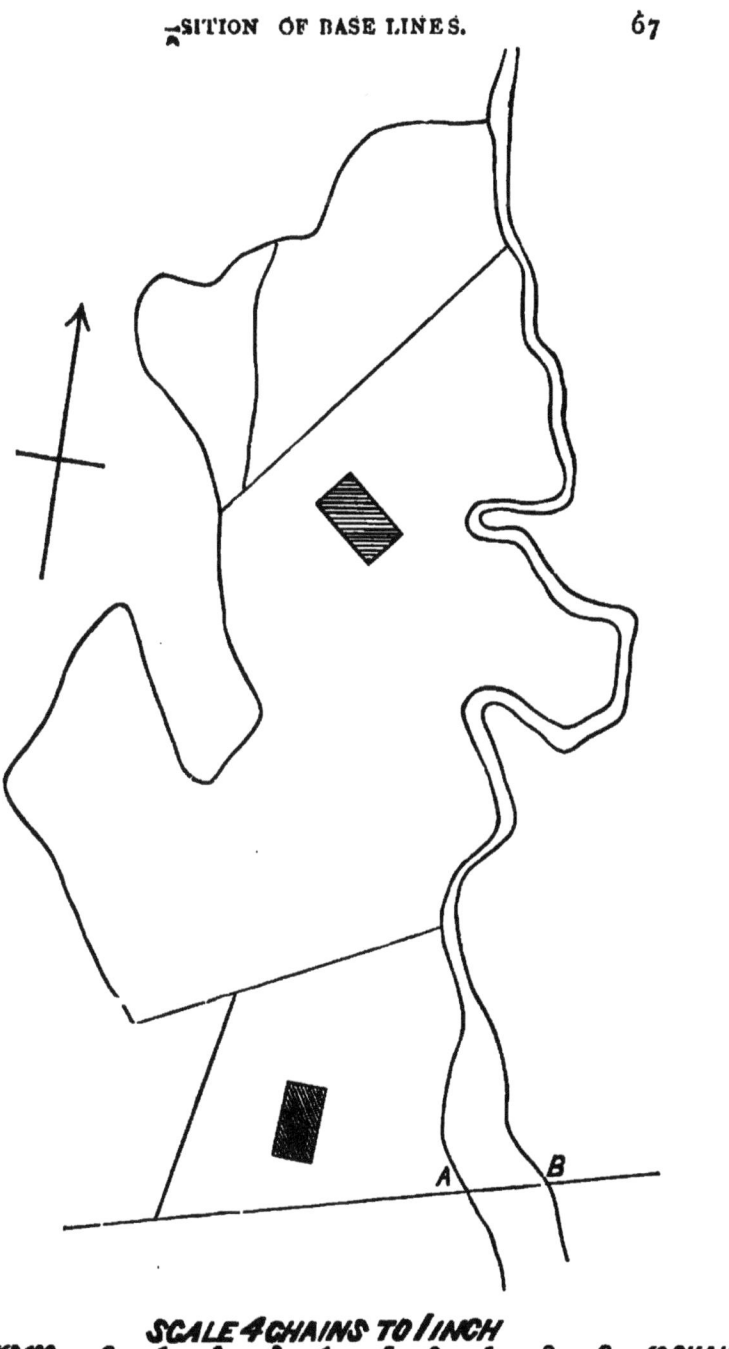

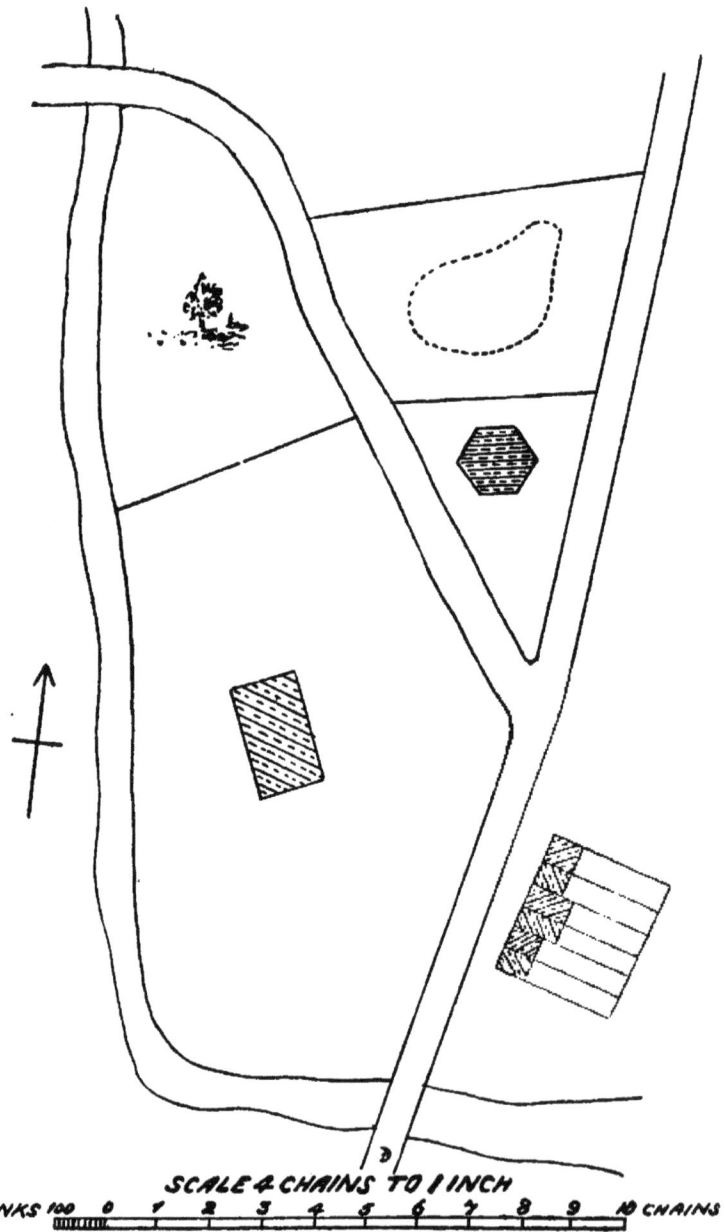

POSITION OF BASE LINES. 67

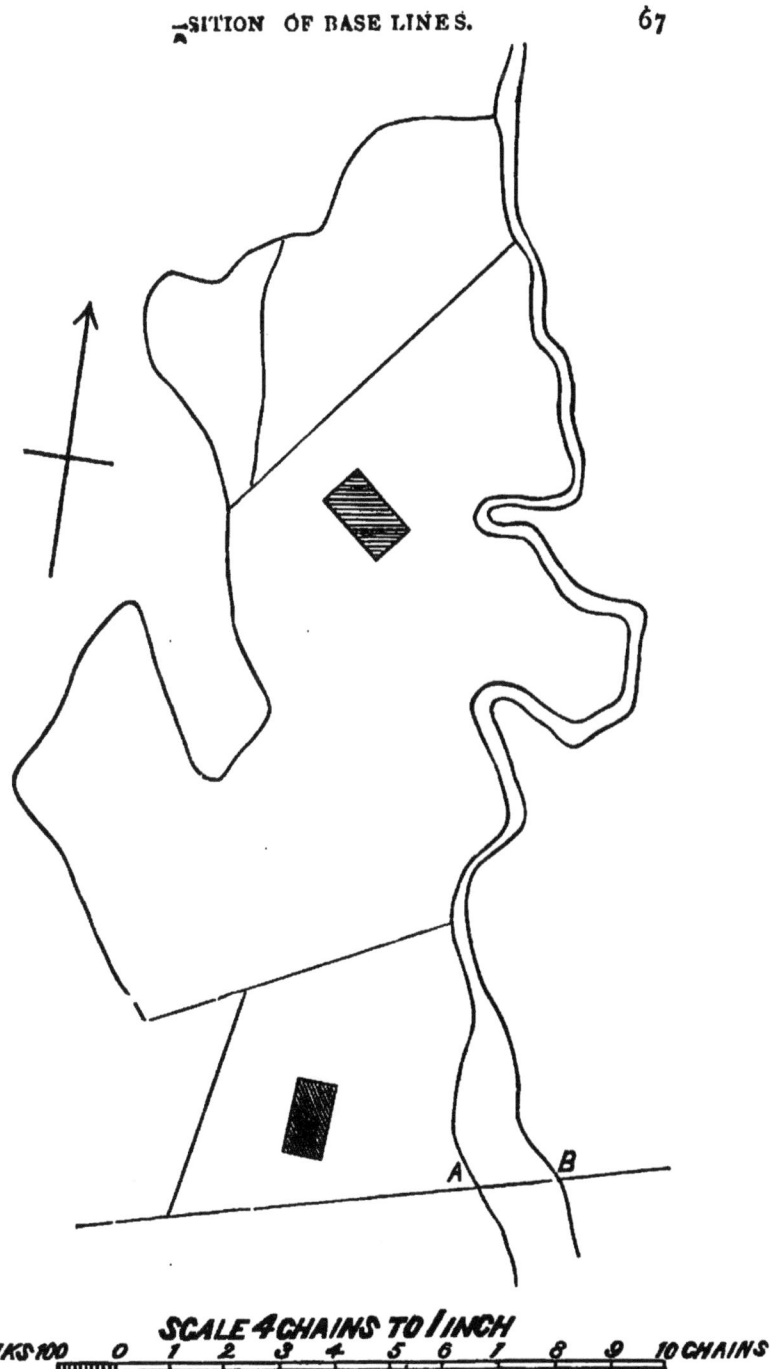

SCALE 4 CHAINS TO 1 INCH

68 LAND SURVEYING AND LEVELLING.

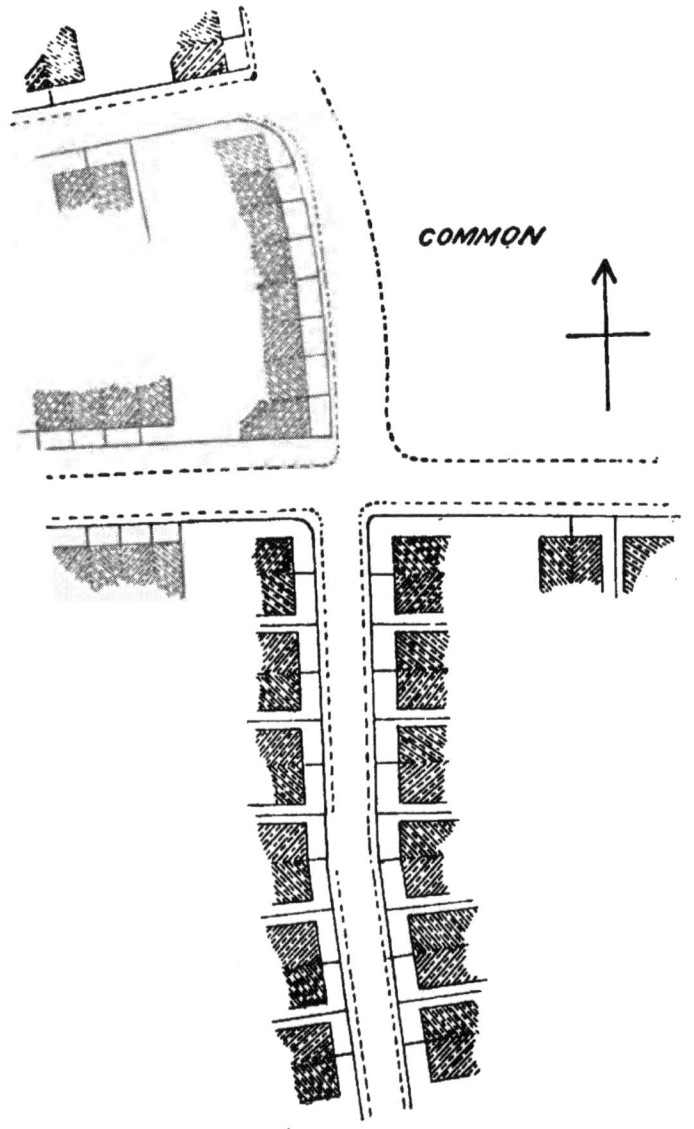

SCALE 4 CHAINS TO 1 INCH

POSITION OF BASE LINES. 69

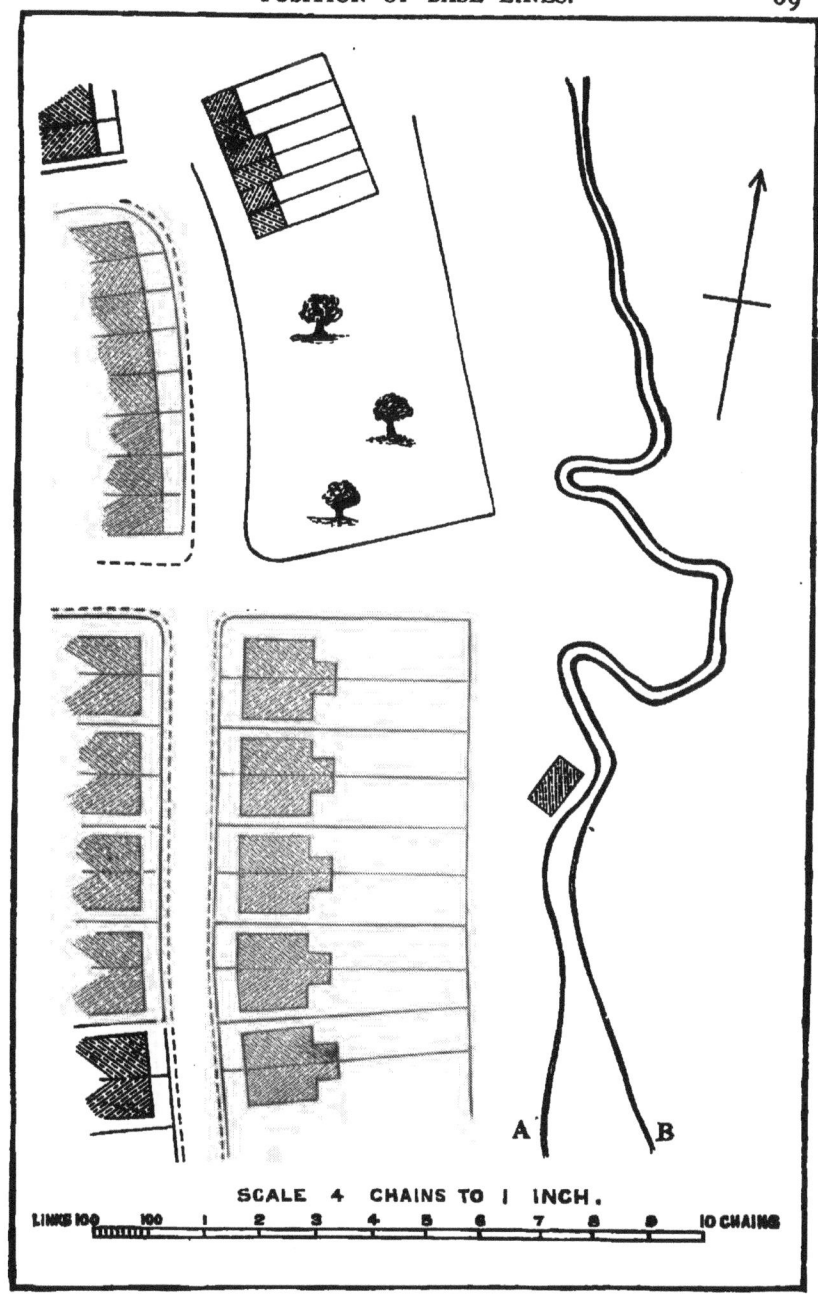

SCALE 4 CHAINS TO 1 INCH.

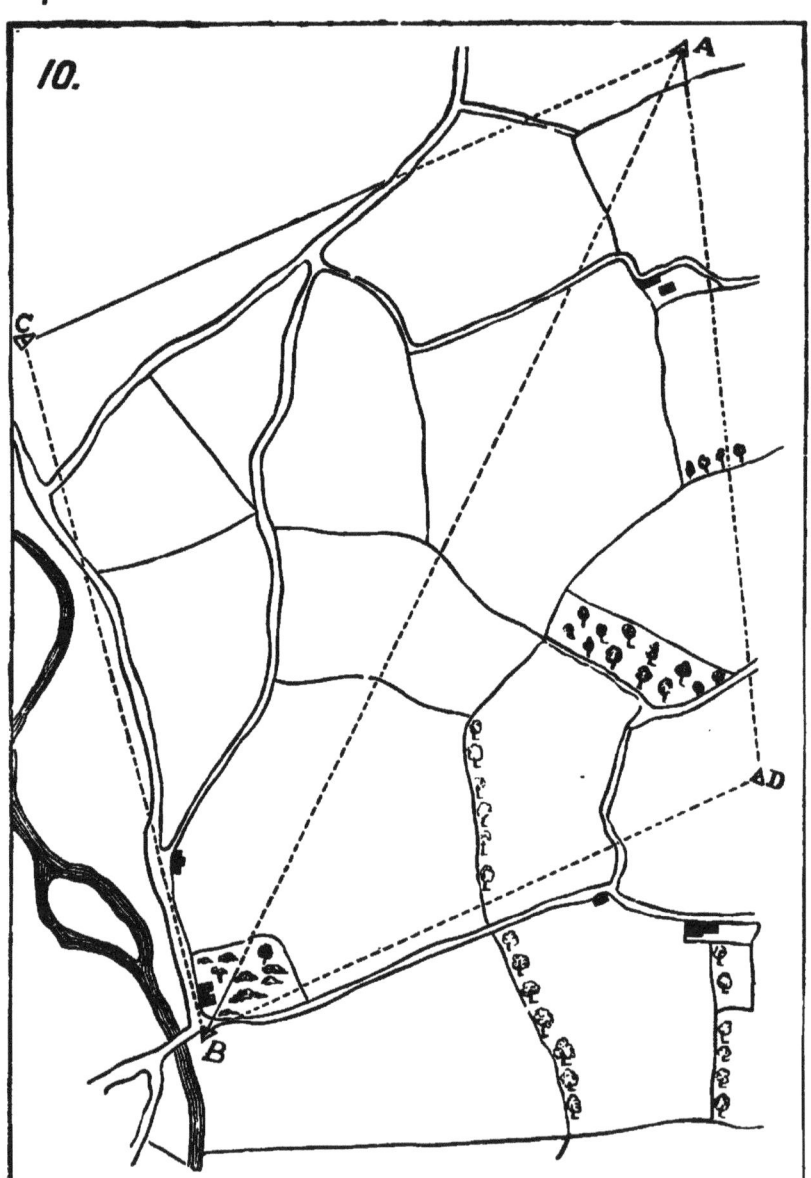

Given a plan in which the base lines CA. AD. DB. BC and AB are run and chained, indicate what other base lines you would run to complete the survey.

telescope when moved in the direction of the hands of a watch. Hence, when the base line to which you are approaching deviates to the right-hand of the surveyor as he walks over the previous base line, the angle booked exceeds 180°, and when it deviates to his left-hand it reads less than 180°. (See page 85.) It will be observed that as the primary scale of divisions is marked in a continuous circle upon the outer edge of the lower plate, a horizontal angle may be repeated indefinitely round the limb of the instrument, The principle of the instrument requires that the graduated circles should be concentric with the axis on which they turn, and also with one another. Errors due to eccentricity may be neutralised by reading the index of two or more verniers, and taking the mean of the number of sub-divisions recorded, because we know by geometry that if two straight lines cutting off opposite arcs of a circle intersect one another in any point within that circle, the sum of the opposite angles is the same as if the line passed accurately through the true centre of the circle. The error of graduation is non-variable in the same instrument, but can be reduced to an inappreciable value by a number of repetitions. The process of repetition is thus conducted :—The lower plate containing the primary scale of divisions being clamped, the zero point marked 360° remains fixed, and the required angle is registered by the vernier F in fig. 2 traversing to G in fig. 5 (pages 54—56) of the diagram "Plan showing base lines." The clamp to the vernier plate being tightened, the line of collimation is accurately set in the direction of the base line B C by the upper milled-headed tangent screw, and the angle A B C in fig. 1 is then read by means of the magnifying eyepiece attached to the vernier at G. To check the reading of the subdivisions upon the vernier at G, the vernier at H may be read. Next let the vernier plate remain clamped, so that after the first reading the upper plate is fixed to the lower; then, without separating these two plates, unclamp the collar to the lower plate and revolve the telescope bodily round to the first line sighted, and set the line of collimation in the direction of the base line B A, by again clamping the collar and working the lower tangent screw. Unclamp the vernier plate and revolve it round in the direction of the hands of a watch towards the direction of the base line B C. Having done this, again set the telescope accurately in position by means of the clamp and tangent screw to the vernier plate, and read the angle. Let us suppose that now the angle reads 196° 20', it will be seen that, adding the value of a complete revolution, or 360°, to this reading, half this total amount would be 278° 10'. At a third reading the same operation should be gone through in a similar manner, and the angle should be within the value of one subdivision read upon the vernier plate of three times the amount of the first angle read, as it will be found in practice that the second and third operations will give a result slightly varying from the first angle, but that a number of repetitions added mechanically serve to secure accuracy in the resulting mean. The advantage of recording the first reading is that it serves as a check upon the number of degrees and minutes in the final average. In rough country it is advisable to commence and terminate the base lines on high and rising ground in every instance where practicable, for the purpose of being better able to obtain a good bearing on the back line when taking the angle of the forward line, and to see if possible the station points or bottom of the poles at the end of the lines forming the angles to be observed. (See page 10.) Winding roads should be traversed with the smallest possible number of angular observations.

The student is recommended to make a sketch of each of the five examples given on pages 66 to 70, each supposed to be drawn to a scale of 4 chains to 1 inch, and to indicate thereon in dotted fine lines the position of such base lines as he would consider necessary with a view to being able to plot each survey from such base lines in the most advantageous manner, also to number the lines so laid down, in order to indicate the sequence in which he would proceed to measure them, then to show the result to some experienced surveyor and seek his opinion thereon. The answers he will receive in reply to his efforts will probably teach him more than mere observations of several examples presented already solved to his eye, and will greatly assist his future judgment in the field.

CHAPTER VIII.

PLOTTING ANGLES.

PROTRACTORS are made into various shapes and sizes; the circular forms are the most accurate. The plain circular protractor has its outer circumference bevelled off to a thin edge, and is marked with radial lines indicating degrees and subdivisions of a degree, which are marked off upon paper with a needle point when required for purposes of plotting. The centre of the circle at which the angle is situated is marked upon the centre of the edge of a fixed bar, which crosses the hollow inner space in the instrument, and which is so constructed that the bevelled edge of this bar gives a true line through the centre of the protractor joining the zero point upon the outside edge with the division indicating 180°. In some instruments there is no true centre, but a small circle around it, with the cross lines leading up to the inner metal circle of the instrument, upon the supposition that the eye will detect a point in the centre of a circle with precision. In most circular protractors a revolving arm is fixed so as to project beyond the circumference. One edge of this arm forms a radial line from the centre of the circle, and, when a vernier scale is attached, the arrow or zero point upon the vernier comes upon this radial line. In the best instruments folding arms are attached as shown in fig. 1, and also to a reduced scale in fig. 2. (See pages 77 to 79.) These folding arms can be fixed in any required position by means of the clamp screw marked D. The complete circle is here graduated into degrees and half-degrees, the verniers read to minutes, and the circle is crossed by four radial bars having the metal removed in the centre. The folding arms and apparatus connected with them, to which the verniers are attached, are fixed upon a frame, the whole of which can be made to move, when the clamp D is slackened round the centre of the instrument. The detail of the attachment of the folded arms is shown in the section A A. It will be observed that, instead of the projecting arm lying flat upon the paper, as in the plain form of instruments, a sharp point, marked H, is fixed to the end of each arm, which, by slightly downward pressure with the hand, will leave a fine puncture upon the paper, and the line joining the opposite points so marked by each arm will pass through the centre of the circle, and will correspond in a line with the reading of the instrument, and thus the angle at the centre can be more accurately set out than could be done by drawing a line along the edge of a radial arm. Hence those instruments fitted with open double arms are preferable to those having a single solid arm in which one edge of the outside portion forms a radial line from the centre. The steel spring shown in the section A A at the joint of the folding arm with the vernier frame enables the point H to become automatically lifted off, clear of the drawing-paper, directly the required position has been marked and the pressure of the hand above it is taken off. The short marks K upon the inner circumference of the circle are engraved upon bevelled edges, as shown in fig. 1 and the section B B. These are placed so that lines joining them cross one another at right angles through the centre of the instrument. The centre of the instrument is sometimes made of glass with fine lines across it in the direction of the lines K K, and the centre is sometimes engraved upon the bevelled edge of a semicircular piece as indicated at B. In order that the line drawn through the points H H passing through the centre of the instrument may cross the two opposite marks K K at zero and 180°, and be at right angles to the line joining the marks K K at 90° and 270°, when the verniers read zero and 180° respectively, the zero points upon the vernier scales are placed in the centre of each scale.

To set the verniers to any required angle, say 145° 33', for the purpose of plotting the angle A B C: having determined the exact position of the station B by scaling its length from the station A, in the given direction A B, the line A B is temporarily extended in length by ruling a

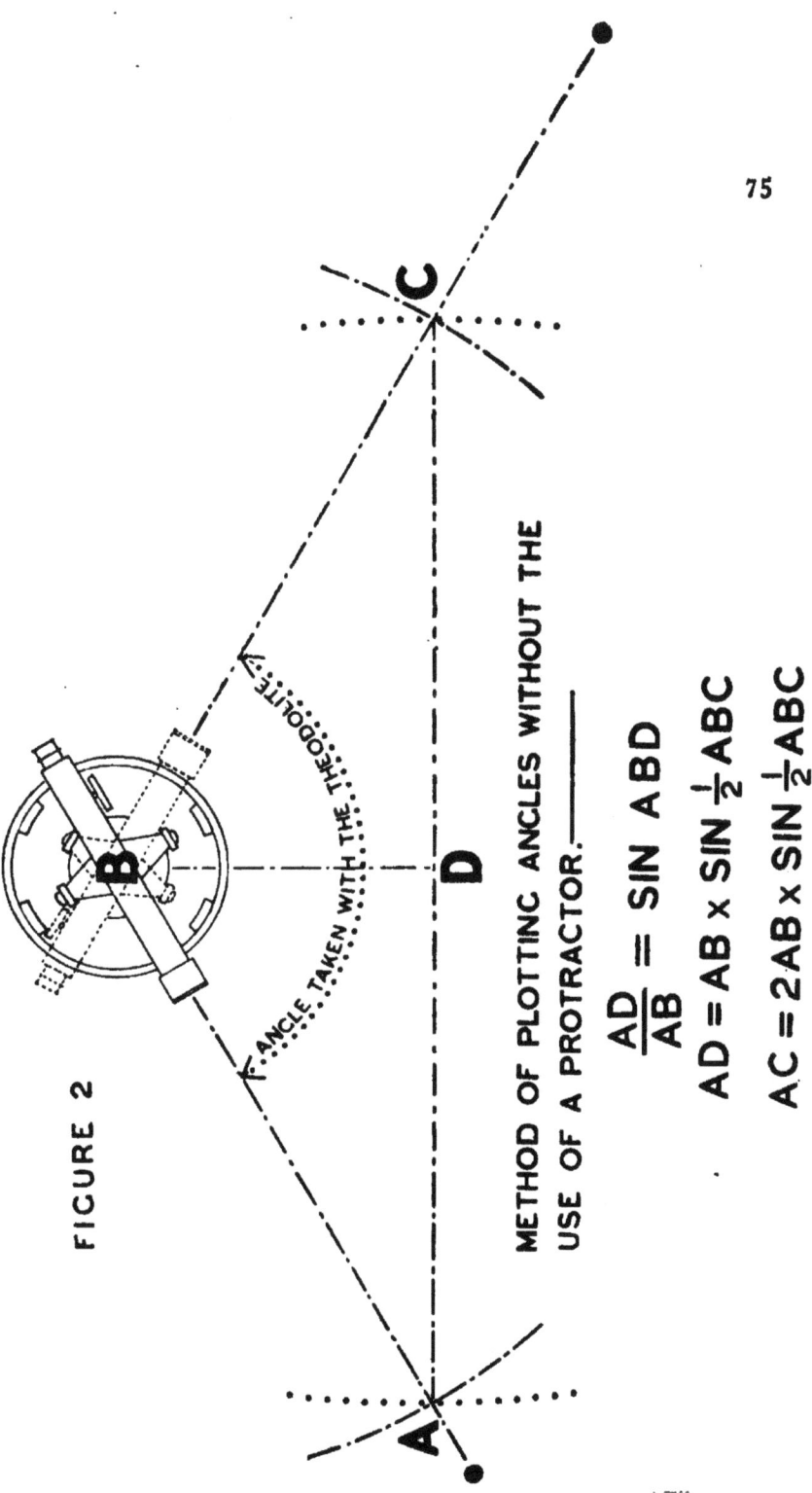

FIGURE 2

METHOD OF PLOTTING ANGLES WITHOUT THE USE OF A PROTRACTOR.

$$\frac{AD}{AB} = \text{SIN ABD}$$

$$AD = AB \times \text{SIN} \tfrac{1}{2} ABC$$

$$AC = 2AB \times \text{SIN} \tfrac{1}{2} ABC$$

pencil line towards the 180° mark, and another pencil line at right angles to it of sufficient length that the marks K K may be each set upon these lines when the centre of the instrument is placed over B and the mark K at zero comes on the base-line A B. The eye can more readily judge of the continuity or non-continuity of a line than it can measure the distance apart of small divisions by observation. Hence the addition of the vernier scale is employed for determining the fractional equivalent of a division upon the primary scale at which the index stands, whereby the eye determines where the line looks continuous, and thus reads into a vernier for arriving at the sub-divisions of a degree. The projecting arms are then opened and travelled round with the hand, by moving the frame at or near D, until the arrow upon the vernier coincides with the *nearest* division upon the primary scale to the reading required. In the present case this would be $145\frac{1}{2}°$, as shown in fig. 4. The clamp D is then fixed, and the vernier is set to 145° 33' by means of the tangent screw. (See fig. 1.) The actual position required is indicated in fig. 3. It should be remembered that the tangent screw is only intended to be applied for adjusting the sub-divisions upon the vernier scales to any *addition* required, which cannot be accurately set by eye upon the primary scale marked upon the continuous circle of divisions. With the use of a protractor as shown in fig. 1, the punctures on the drawing made by the sharp index point near the end of each arm, exactly correspond with the reading of the instrument, and it is an advantage for the surveyor to employ a circular protractor that shall be divided exactly similarly to the primary circle and vernier of his theodolite. (See pages 77 to 79).

Fig. 8 (pages 99 to 101) illustrates the plotting of the angles taken between the base-lines in a traverse survey by means of the circular protractor. The station pegs which are indicated in the field, as shown by figs. 10 and 11, are marked upon the plan with a small pencil circle, surrounding the station point, as shown at the termination of the base-lines Nos. 1 and 5, in fig. 8 of this diagram.

Fig. 7 (pages 99—101) illustrates Stanley's patent protractor, the construction being explained by its accompanying

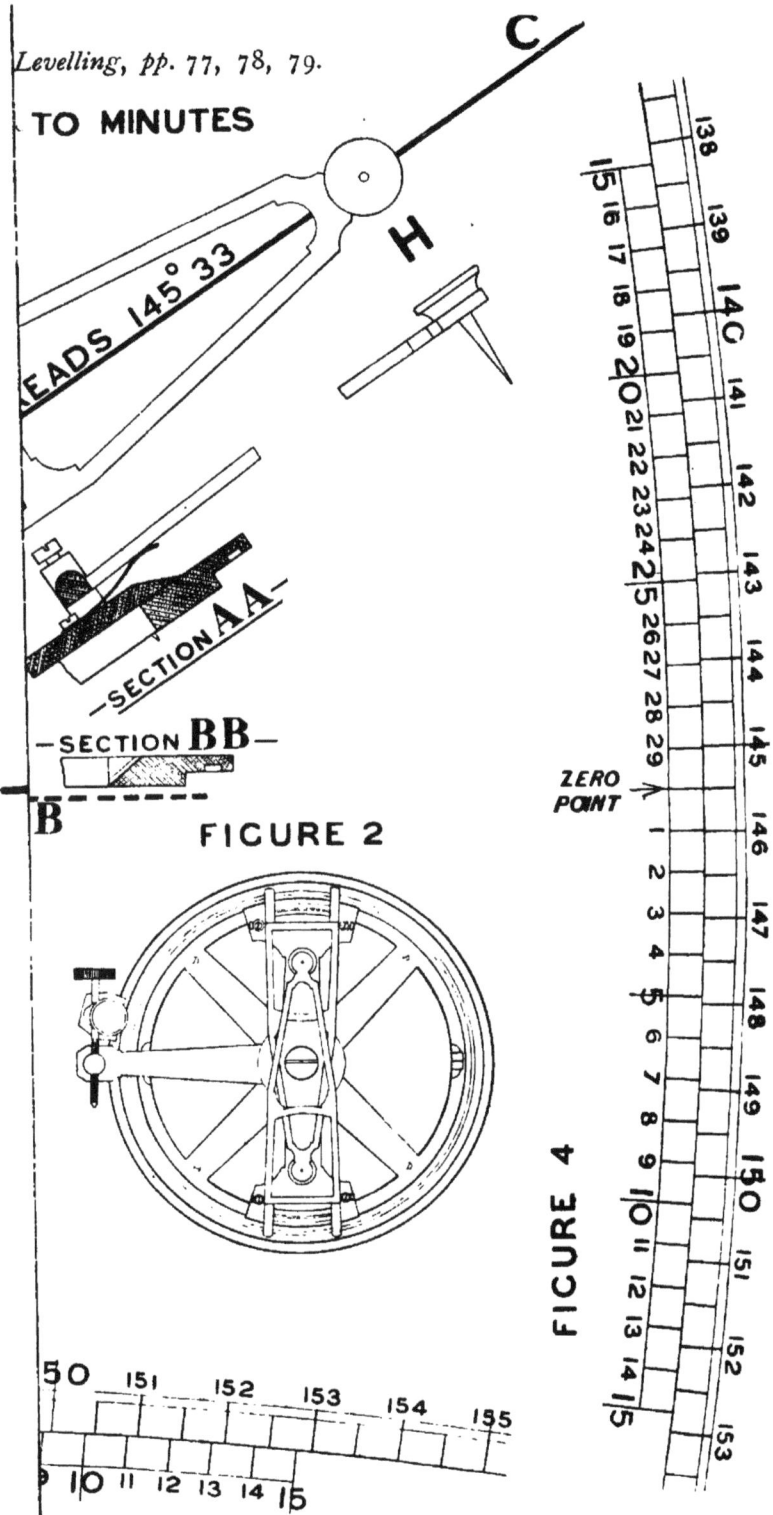

Levelling, pp. 77, 78, 79.

FIGURE 2

FIGURE 4

sections. By its use angles already plotted upon paper can be measured off without drawing in extra lines upon the map or plan. The instrument is formed of two concentric circular pieces. The outer of these circles is divided into degrees, and sub-divided either into halves or thirds of a degree, while the inner circle carries a vernier, which enables the operator to read to minutes. Two arms, as shown in plan and section, are mounted upon these two circular rings, and have each one of their edges radiating from the centre of the instrument. These arms can be moved to include any angle up to nearly 360°, the amount or size of which is then recorded by the index mark upon the scale of divisions. The instrument is made of metal, having the underside of the two radiating arms and the lower surface of the outer ring in the same plane, so that when the protractor is placed upon the plan the arms can be set to the lines between which the required angle is to be measured. If these lines do not actually meet, the length of the arm permits the angle to be set, without producing the direction of the lines in pencil to determine their point of intersection. For use in hot climates, protractors are also made of vulcanite, with vulcanite arms or blades, the material being annealed at a very high temperature, so that when subjected to the severe test of the extreme heat of India, the temperature shall not influence its form to an appreciable extent.

Fig. 9 (page 81) illustrates a station pointer, which is used in a similar manner to a protractor, but it will be observed there are two radiating movable arms with verniers attached, one upon each side of a central arm fixed at zero. These arms, marked C, are each set by means of their respective clamp and tangent screw at D, to the required angle with the line B A. The instrument thus clamped is moved about upon the plan, until the edges of the fixed arms B C, B A, and B C respectively coincide with the points upon the plan which indicate the stations observed, when taking the two angles with B A in the field. By means of a needle pointer (fig. 14, pages 99—101) the position of B can then be fixed. Lengthening bars to the arms are provided with the instrument for use when required, jointed as shown on page 81.

The most accurate way of plotting angles is by the intersection of arcs, having calculated, as shown in figure 2 (page 75), the length of an hypothenuse to a triangle the sides of which are taken of equal length. The triangle

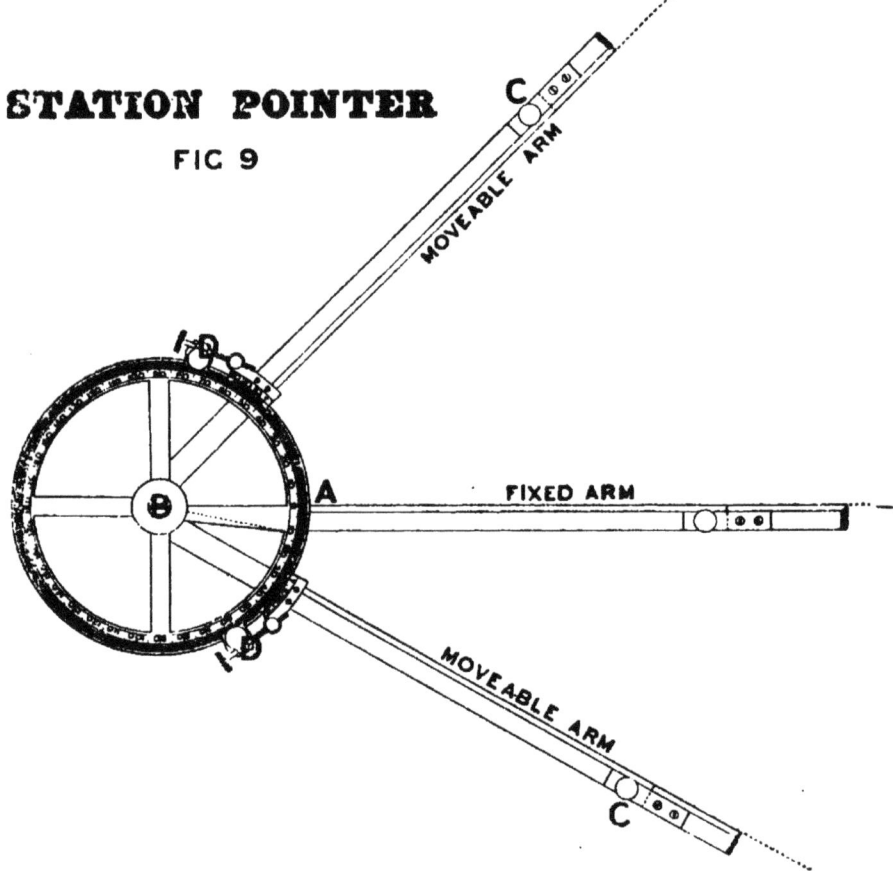

STATION POINTER

FIG 9

A B D will afford the greatest probability of accuracy for calculation, when, as in the figure A B D, being less than a right angle, and the angles adjacent to the base are nearly equal, but even small acute and large obtuse angles may be most accurately determined by means of their trigonometrical values. It is better to plot the triangles of a survey from the calculated lengths of their sides than by measures of angles, because measures of length can be taken from a scale and marked on a plan more accurately than the direction of lines pricked off from a protractor can be determined. (See Chapter IX.) For rough purposes the scale of chords found upon the short plain scale which accompanies nearly every box of instruments is sufficient. The application of this scale, as shown below, depends upon the

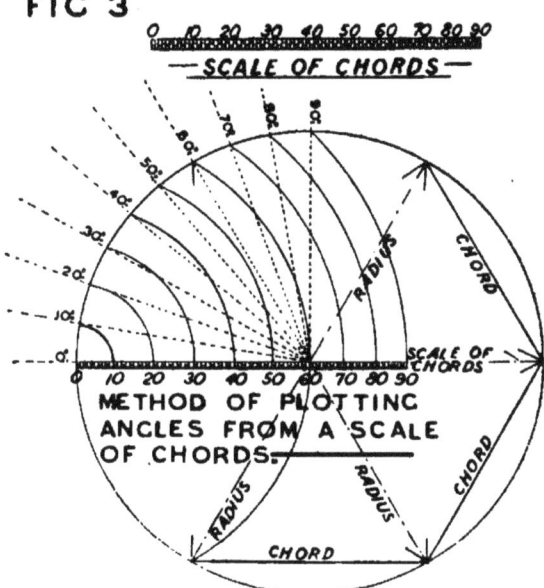

FIG 3

METHOD OF PLOTTING ANGLES FROM A SCALE OF CHORDS.

fact that the chord which subtends an angle of 60° at the centre of the circle is equal to the radius, so that by describing a circle with a radius equal to the chord of 60° other angles can be set out by intersecting this circle with a chord equal in length to that given upon the scale for the required angle, and drawing the angle at the centre of the circle which subtends that chord (fig. 3). But for important surveys either a circular protractor having a primary scale of divisions and verniers similarly divided to the theodolite may be adopted, or the application of trigonometry introduced, as explained in the chapter on "Transverse Surveying." In comparing a diagonal scale with a vernier scale, the student has only to draw out one of each for himself to arrive at a satisfactory conclusion. To form a diagonal scale, draw lines parallel to the line required to be sub-divided, at equal distances apart, say between C and D, and join the opposite corners by a diagonal line D B. The points of intersection of this diagonal line D B will then mark upon the parallel lines successively the fractional parts of the length of line C B, and by drawing vertical lines from these points of intersection the respective lengths may be transferred to the line C B. Where the sub-divisions are constructed in tenths, as is usual upon the broad plain scale of six inches in length usually found in a box of instruments, the decimal parts of any length may easily be measured by means of a pair of compasses with the aid of a diagonal scale, but it will be seen that a vernier scale is preferable, because in a diagonal scale it is not so easy to draw the diagonals with accuracy between the first and second lines, and there is no check on its errors, whereas the vernier scale would more readily display any inaccurate construction. Hence the use of the vernier scale, as explained upon pages 76, and 77 to 79.

CHAPTER IX.

TRAVERSE SURVEY.

The term "traverse," as used and applied to surveys, denotes a continuous line in a zig-zag direction, the amount of inclination of the base lines with one another composing the traverse being taken carefully. When, as in an extensive survey, the station points to be plotted are numerous, it is found preferable to work upon a co-ordinate principle with the aid of trigonometry. A straight line 360° to 180°, or ranging north to south, is drawn upon the paper in approximately the centre of the plot, and a point of origin being assumed as a centre, the space surrounding it is next divided into squares, the sides of which are accurately drawn to run both north and south, and west and east of this point of origin, thus treating the earth as a globe having many faced solids, each comparatively so small that any difference in the continuity of area of adjoining faces is negligible. The angle of each base line is taken in the field with the direction of north by the aid of a theodolite, but instead of being determined at each station by the swing of the magnetic needle, the inclination to each other of the base lines is taken and reduced in the following manner. Columns for entry may be ruled in a field-book with the following headings, thus (see diagram, page 85):—

Line.	Observed angle.	Whole circle bearing.	Length.	N.	S.	E.	W.	Totals.			
								N.	S.	E.	W.
0–1	78°	60°	100	50	...	86	...	...	...	...	...
1–2	255°	135°	150	...	105	105	...	...	55	191	...
2–3	75°	30°	120	103	...	60	...	53	...	251	...

LAND SURVEYING AND LEVELLING. 85

TRAVERSE SURVEYING.

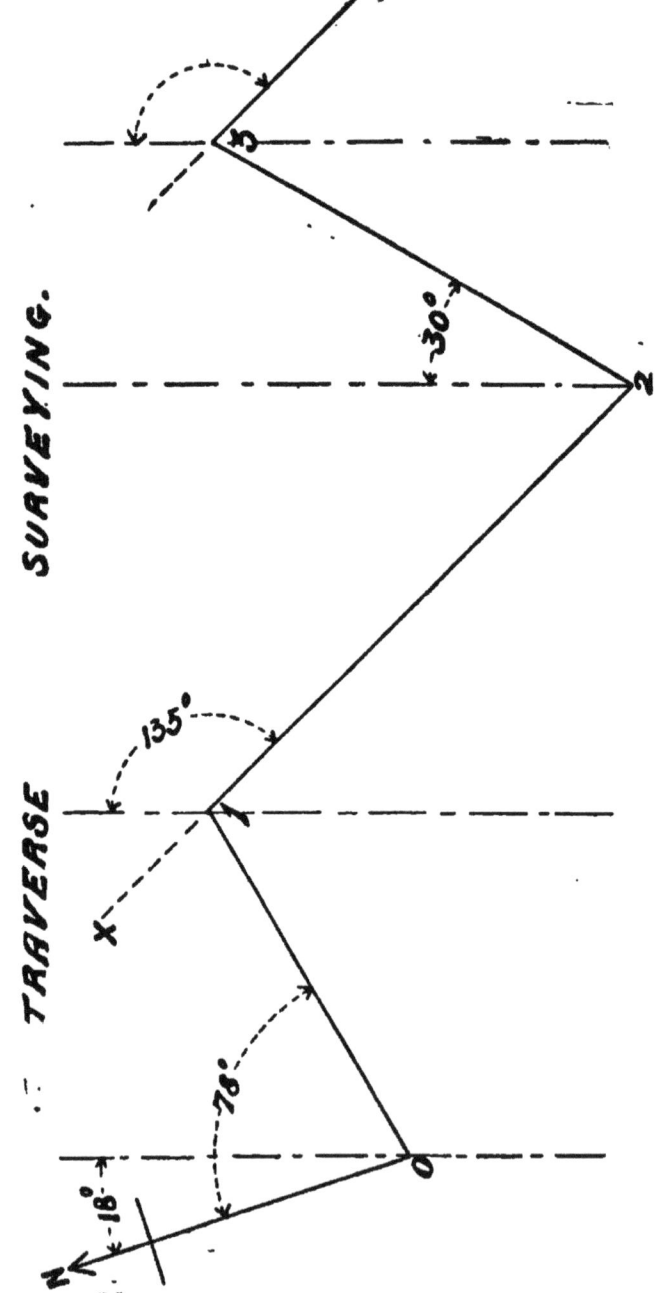

But this form of diagram is not arbitrary. The surveyor may arrange his own form of entry, and also decide as to whether he will read exterior or interior angles, that is, whether he will proceed east to west viâ south, or east to west viâ north. Supposing he commences at the north of a circle and travels round in the direction of the hands of a watch, this being the manner in which a theodolite is usually graduated, he will record exterior angles. Referring to the annexed figure, his first reading is, of course, the angle which the first line taken, namely, line 0—1, makes with the magnetic needle when at rest, say 78°. This must be most carefully observed and recorded, but the magnetic north being, say, 18° west of the true north, the angle which the line makes with the direction of the true north will evidently be 60° east. This is called the whole circle bearing. Setting up the instrument over the next station the angle between lines 0—1 and 1—2 would read 255°, and subtracting therefrom 120° we obtain the whole circle bearing of 135°. From this it will be evident that if the surveyor clamps the upper plate of the instrument to 240° or to the sum of the last angle 60° with 180°, and then sets the telescope upon the line 1—0, the zero of the primary scale with the lower plate thus clamped will be coincident with the north, and when the telescope with its upper plate is traversed round to the direction of the line 1—2, the vernier will record the required angle or 135°. Again, suppose the observed angle between the lines 1—2 and 2—3 to be 75°, the whole circle bearing is found by subtracting 45°. This is a tedious process and requires a sketch made at each station, whereas if the surveyor clamps the upper plate of the theodolite to 315° or to the sum of the last angle with 180°, namely 135° plus 180°, and then sets the telescope upon the direction of the line 2—1, the zero of the primary scale when the lower plate is thus clamped, will be coincident with the north, so that when the telescope with its upper plate is traversed round in the direction of the line 2—3, the vernier will record the whole circle bearing of 30° east of north. It will be evident that facility is afforded by the use of a theodolite with the usual two verniers diametrically opposite to one another whereby a difference of 180° exists between their respective readings, so that if when the instrument is set up over station 1, the vernier plate is clamped to the angle furnished at the previous station with the true north or 60°, that when the lower plate is clamped and the upper plate carrying the telescope traversed round to view in the direction of the line 1—2, if we read the opposite vernier or the angle 1 to X, we should record the angle 135°. The opposite verniers may be distinguished by being right and left of the telescope with its bubble up or bubble down, or by placing a magnifying eye-piece over the vernier in use, and not over the other vernier. Again, when viewing at station 2 in the line 2—1, and fixing the vernier at 135°, then clamping the lower plate in this position, when the telescope is traversed round to the direction 2—3 it would record 210°, but the opposite vernier would read 30°, which is the angle to be recorded. The northings and southings with reference to the point of origin are termed differences of latitude, while the departures left and right are termed eastings or westings according to the direction in which they are taken. The calculation of the differences of latitude and departure is effected by reference to trigonometrical tables (Chapter V.), and thus by an application of rectangular co-ordinates referred to a four-quadrant system, the plotting is found to be more accurately determined than when the method shown in fig. 8 (pages 99—101), by means of a circular protractor, is relied upon. Each square upon the sheet of paper employed for the plotting is drawn to have a length of side calculated to scale a fixed round number of links, not only to aid rapid plotting but to facilitate final accuracy. If the survey was required to be drawn to the Ordnance scale of $\frac{1}{2500}$, then squares each one-tenth of a foot in the side would represent 250 feet lineal measurement, an amount which needs no consideration of any allowance to be made between the spherical and plane surface of the ground traversed. If the length of line 0—1 be 100 links, then the station 1 will be situated at 100 cos 60° to the north and 100 sine 60° to the east. If the length of line 1—2 be 150 links, then station 2 will be 150 cos 45° to the south and 150 sine 45°

to the east, so also if the length of line 2—3 be 120 links, then station 3 will be 150 cos 30° to the north and 150 sine 30° to the east. In this way the co-ordinate measurements north, east, south and west of the point of origin are determined. The length of base lines would be usually greater than the number of links here assumed, and the angles would generally contain minutes in addition to degrees, but the method would be the same, and the student is recommended to try a case by traversing an imaginary enclosure of five or more sides pegged out in a field and to frame his own mode of entry in the field-book.

The figure (page 85) illustrates a portion of an unclosed traverse, in which it is presumed that the surveyor decides from real cause of situation, or of option, not to work back to the starting-point, but in all important surveys it is usual to finish upon the starting-point, even when necessary to pursue a circuitous route for this purpose, as the summation of the angles can then be checked by the well-known corollary to the 32nd proposition of the first book of Euclid, which states that the summation of all the interior angles of any rectilineal figure, together with four right angles, are together equal to twice as many right angles as the figure has sides. In a traverse survey the number of stations should be as few as possible, and as much care should be exercised in taking the lineal as in taking the angular measurements, but it is immaterial whether the angles are read before or after the chaining is done. It is nevertheless found practically that the length of a very long line upon uneven ground can be more accurately determined trigonometrically than by the direct application of any measure of length, and in cases of surveys, in which the distances are long and numerous and the offsets comparatively few, this method is suitable. To secure great accuracy in taking important angles between very distant stations the best time for observation is about an hour before sunset in dry weather, and early in the morning, if not misty, in wet weather. It is advisable to avoid the middle or close of a hot day in damp weather, if possible. If the surveyor is satisfied that the graduations of his theodolite are correctly marked he need not read more than one vernier; but should he

F

decide to read the opposite vernier, as a check upon his record of sub-divisions, it is a safe plan to have some means of designating the verniers apart. Some makers engrave the letter A upon one, B upon the other, and C upon a third if provided. When not so lettered or marked, the surveyor may use only a single reading lens which travels over the vernier to be read and he thus avoids doubt. Col. Everest's theodolite usually has three verniers, while the ordinary plain and transit instruments usually have only two.

CHAPTER X.

PLOTTING A PLAN.

MAPS and sea charts were first introduced to England in 1489 by Bartholomew Columbus, brother to Christopher Columbus. A geographical map indicates the relative position of places. A topographical map gives a plan of all the principal points and objects. The former corresponds to a key plan, the latter to the plan of a survey, which gives a picture of the site such as would appear from a balloon floating in the air exactly over every point of the area viewed.

The term "plotting" refers entirely to the office work—laying down the base lines and filling in the survey. In chain surveying the accurate position of base lines from which offsets are taken depends upon the intersection of measured distances. In fig. 1 (pages 92-94) the base lines A B, B C, C A, measure respectively 14 chains 88 links, 23 chains 12 links, and 32 chains 74 links. The plan need not necessarily be plotted so that the top of the paper represents the direct north, as in an ordinary map; but the lines should be placed in the most convenient position to suit the shape of the survey. It will generally be found best to commence by drawing the longest base line with the aid of a straight edge, or if the straight edge be of insufficient length and the paper free from buckle, a good substitute will be found in a thread blackened with a burnt cork and snapped when correctly stretched over the points, or the line may be determined by carefully puncturing the drawing paper in several places by marking fine needle points on the thread line and afterwards joining them, taking care when drawing the line through the needle points that at least half the length of the straight edge be applied to the portion already drawn as a bearing by which the additional length may be accurately continued in a straight direction through the needle points. Proceeding in the same order as that followed in the field, the position of the stations A and C, at a distance of 3,274 links (see fig. 1) apart, are then marked with a needle pointer (see fig. 14, pages 99-101) upon the line A C, and are temporarily indicated for future reference in the process of plotting by surrounding the station point with a small circle drawn in pencil, as shown at the points marked A and F in fig. 8 (pages 99-101). The position of the station at B is next determined by striking an arc from the station at A with a radius equal to 1,488 links, and then intersecting it with an arc having a radius equal to 2,312 links struck from the station at C. (See pages 92-94.)

Fig. 2 illustrates four different positions of arcs for the determination of station points from chain measurements, those at E and F being clearly defined, while those at G and H overlap one another to such an extent that it is difficult to accurately mark their exact point of intersection. The angle between the base lines should, as near as practicable, be not less than 60° or more than 90°. If the sides of the triangle A B C have been correctly measured in the field, and accurately plotted to scale, the tie line B D drawn from station B to 1,580 links upon the line A C should scale the length measured in the field, namely, 10 chains 51 links. (See pages 92-94; see also page 4.) The straight edge when not in use should be hung up out of the sun or draughts of wind, both having an influence to render a wood straight edge imperfect.

In this country, the inch being the unit of measurement by which the eye is accustomed to judge distances upon paper, it is usual to state the proportion of the plan to the ground as so many divisions to one inch, but in scales of chains, links are usually reckoned the unit of measurement. As the chain used in the field contains 100 links, plotting and offset scales are decimally divided. The subdivisions upon the edge of each scale read simply a certain number of divisions to the inch marked upon the scale. Thus in the scales shown upon the line A C the mark 10 simply means that there are ten equal subdivisions to the inch. If, therefore, the scale to which the plan is to be plotted be a scale of one

Land Surveying and Levelling, pp. 92, 93, 94.

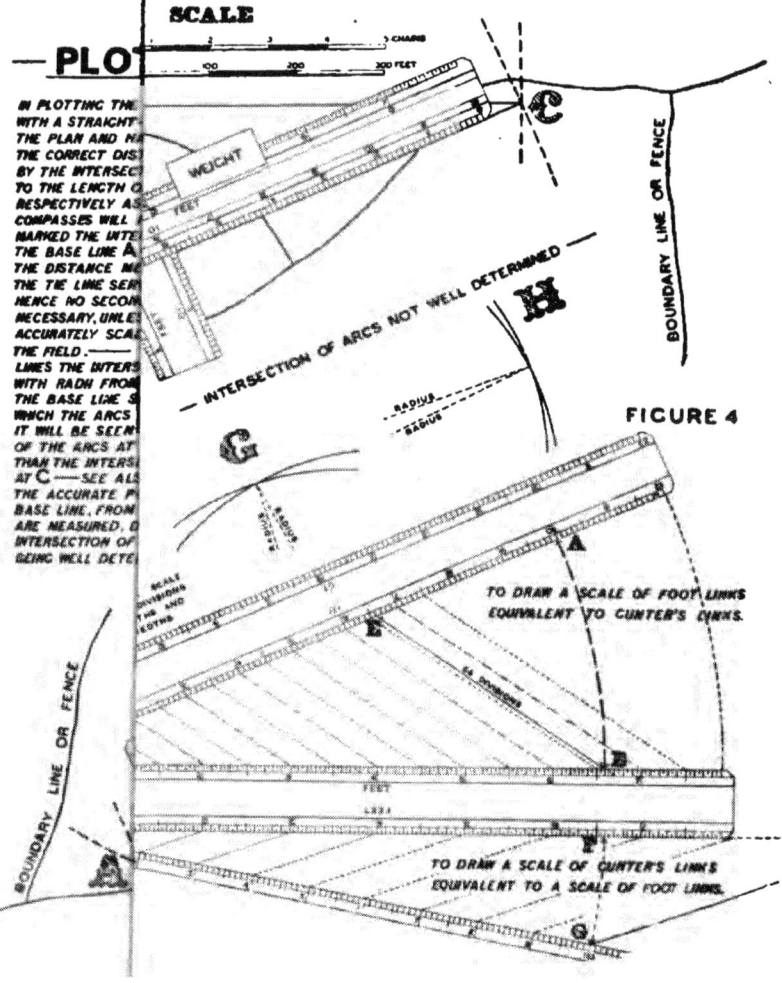

FIGURE 4

chain to the inch, each subdivision upon the scale would represent ten links, and the unit lengths would have to be estimated by the eye. In some scales, divisions marked "feet" are shown upon one edge, which is intended to give the equivalent in foot links corresponding with the scale of Gunter's links marked upon the opposite edge. Upon reference to fig. 3 it will be seen that a length of 3 Gunter's chains or 198 feet (3 × 66 feet) is nearly the same length upon the scale as 2 chains (200 feet) upon the edge marked "feet." If the chain used in the field be the 100-feet or any foot-divided chain, and the plan be plotted to a scale of feet, with the use of the scale of decimal equivalents to an inch, the divisions marked feet upon the opposite edge of the scale would be meaningless; but if the chain used in the field be a chain of Gunter's links, and the plan be plotted to a scale of one chain to an inch, with the use of a scale containing ten or more divisions to the inch, then measurements scaled with the edge of the scale indicating feet, will give the equivalent length in foot links. Sometimes it is useful to plot with a scale of "links" equal to a certain number of feet to an inch, when Gunter's chain has been employed in the field, and it is desired that the scale of the plan shall, for future reference, be an exact number of feet to 1 inch. (See pages 92-94.)

Before attempting to plot the fences, let all the chain lines be first plotted, with their number, their length, and their angle (if taken) written against them in pencil for reference until the whole survey is completed, as it will save time, instead of having to frequently refer to the field-book; and with this object in view, it is also well, when scaling long lines, to mark in pencil every ten chains upon the line. The thin edge of the straight ruler is the best for the pencil lines, and the thick edge for inking them in. All survey lines should be first inked in with light blue lines as being not too prominent a colour. Do not plot as a rule from pencil base lines.

Fig. 4 (pages 92-94) shows the method of *graphically* converting a given scale of feet to a required scale of Gunter's links, and also a given scale of Gunter's links to a required scale of feet. Two straight lines, C A and C B, intersecting at C, are drawn at any angle apart. About 30° is a favourable angle to fix upon. The scale of Gunter's links is applied to the line C A and its subdivisions are pointed off upon this line. From C as centre with C A as radius the distance C A is transferred to B. A line is then drawn from sixty-six divisions upon the scale of Gunter's links to the point B, and the line C B, representing the length of one hundred Gunter's links, is then subdivided into sixty-six equal parts representing 66 feet, by drawing lines parallel to E B from the subdivisions upon the line C A to the line C B. Thus a series of similar triangles is formed, and a scale of feet equivalent to the given scale of Gunter's links is arrived at. Conversely, to draw a scale of Gunter's links equivalent to a scale of foot links, sixty-six divisions upon the line D F are transferred to the line D G, which is then subdivided into one hundred parts by drawing lines parallel to a line joining the point G with the length of one hundred divisions from D, measured in the continuation of the line D F.

There are three kinds of scales, those consisting of a line of divisions at equal distances apart upon the edge of a scale, diagonal scales, and vernier scales. Cardboard scales can not be maintained sufficiently accurate for plan work. Feather-edged scales of a flat section in boxwood or ivory are preferable to similar oval section scales. The latter, though easily tipped up and useful for taking single dimensions, lack the steadiness needed for the continued use of a plotting scale. A flat section offers the fullest amount of frictional resistance to slipping over the drawing paper, and its feather edge enables the divisional marks to come in close contact with a base line. There should never be two different scales upon the same plotting scale, as mistakes may thus be invited to occur, but the distinction between the chain and feet edge of the scale is sufficiently manifest to prevent error. The divisions upon the bevelled edge should read in one direction only, and the underside remain plain without marking of any sort, except perhaps the owner's name.

In plotting a large plan, the very first thing to be done is to draw a scale upon the paper, because paper shrinks. The scale so drawn will then lengthen and shorten by change of temperature in the same proportion as the plan itself, and consequently will give more reliable measurements than the scale taken fresh from a box. An inspection of Ordnance maps, as described on pages 124-126, shows that the scales have shrunk with the paper from their true proportion. (See Chapter XII.) Ordinary plotting scales are usually made of boxwood or ivory. Mr. W. F. Stanley, in his useful treatise on "Mathematical Drawing Instruments," states that the boxwood most suitable for the purpose is rather small, live Turkey wood. It should be of a clear yellow colour and of dense waxy grain. Soft, inferior wood soon becomes dirty in use, and the divisions upon it appear woolly. The white, opaque ivory used is principally imported from the eastern coast of Africa and the Cape, and is preferred by many draughtsmen to the green ivory, which comes from the western coast of Africa, being the least expensive, shrinks less, and has the advantage of showing divisions and figures much more clearly than the green ivory. It has, however (adds Mr. Stanley), one defect. It turns yellow after a few years' exposure. Green ivory is very transparent, of a dull, heavy colour; it does not show the divisions very distinctly until it has been some years in wear, when it becomes of a pearly whiteness, which is unchangeable. Vulcanite is too much affected by changes of temperature to serve as a fit material for scales.

In measuring a long line with a scale one foot in length, having decimal divisions, it is necessary to divide the length into sections, and the best method to adopt is to take 10 inches at a time, leaving 2 inches overlap of scale for accuracy of direction. Upon no account should the compasses be allowed to scratch and mark the scale when measuring any distance. Draw a border line and use this straight line upon the paper, or another thinner straight line outside the border line, to apply the scale, and mark off the required distance thereon with a needle pointer. Then extend the ordinary compasses or the beam compasses so as to transfer the required distance thus obtained to any required position, taking care to keep the legs of the compasses perpendicular to the line. The triangles are first plotted, as recommended upon page 95, and then the details. Since it is advisable to avoid acute angles, the interior angles formed by the tie lines should in the field be measured from the apex to the base as near as possible at right angles to the base, and then when plotting these lines their intersection with the main base lines will be clear and defined. Where the angle to be determined is less than $30°$ or more than $150°$, such tie lines are said to be ill-conditioned, and need to be connected by more than one tie line to ensure accuracy.

The station points $\underline{A}$ are marked upon the paper by a needle point surrounded by a temporary circle in pencil for sake of clearness, thus, $\odot$. It should be borne in mind that the surveyor uses the word point to express position only, without any reference to magnitude.

Fig. 5 (pages 99-101) illustrates a good form of drawing table, having a slot or slots, as shown in the section, fig. 6, for the drawing-paper to slide into when the draughtsman leans against the outer rail in the process of plotting. The inner longitudinal rail can be supported to the drawing either by transverse pieces fixed at intervals or wedge-shaped pieces placed about two feet apart. The whole board rests on trestles, or is otherwise supported level at a convenient height from the floor. By the use of this precaution the drawing-paper is prevented from becoming creased when plotting at the top of the plan, as it can be pulled through the slot, as indicated by the dotted line in fig. 6. Speaking generally, a drawing board is seldom used by the land surveyor. A large and perfectly level table, fitted as shown in figs. 5 and 6, pages 99-101, is more convenient. For small surveys, however, a drawing board may be useful, as it is easily twisted about, over a table to suit different directions of base lines when plotting the offsets. When called upon to supply engineering or architectural details, a drawing board and tee square become almost indispensable to a surveyor. Mr. W. F. Stanley, whose drawing boards excel those of most other makers, describes a good drawing board in his "Treatise" above referred to, as a very great desideratum to the draughtsman. "The qualities it is important

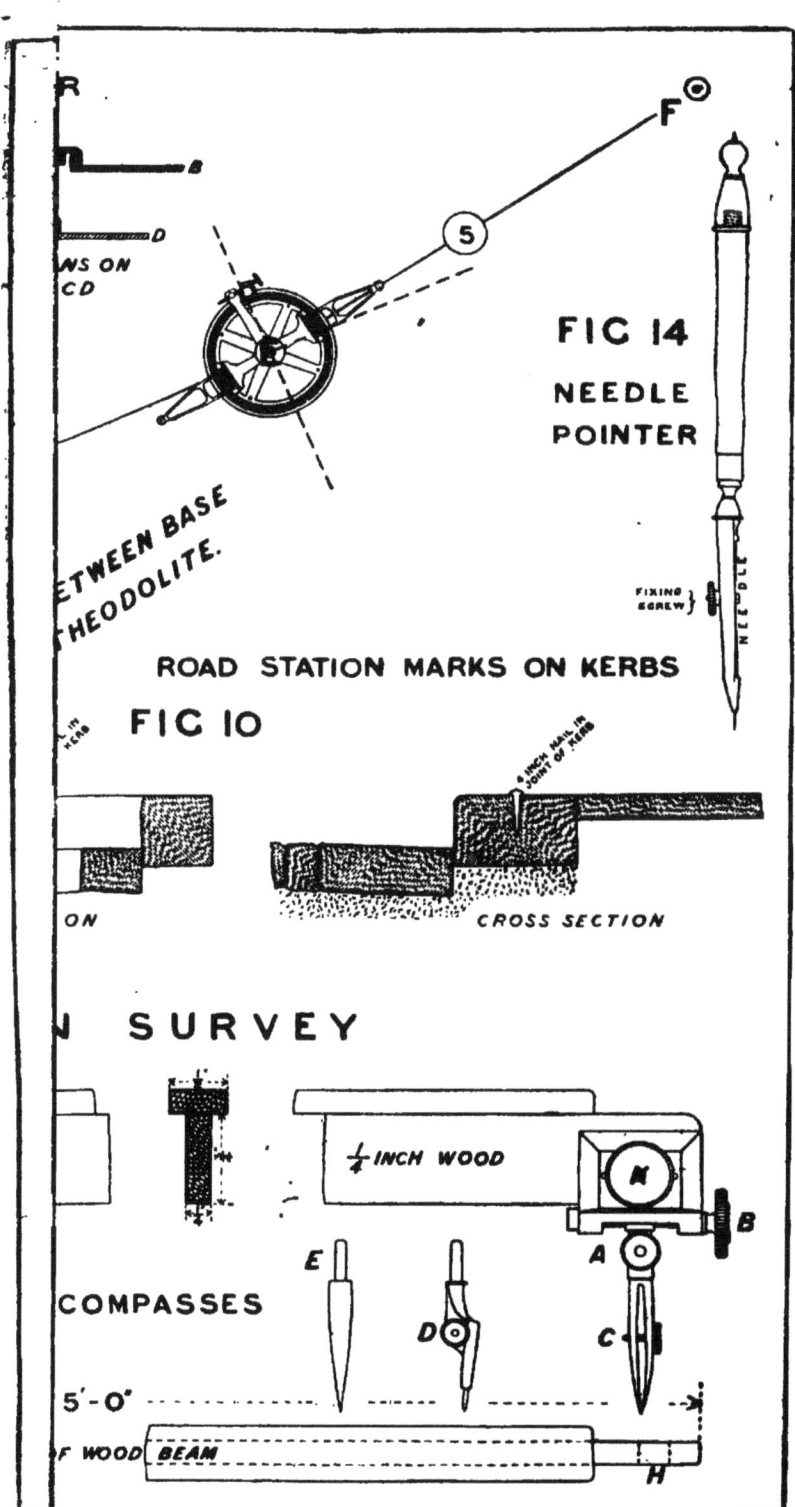

that it should possess are an equal surface, which should be slightly rounded from the edges to the centre, in order that the drawing paper when stretched upon it may present a solid surface, and that the edges should be perfectly straight and at right angles to each other. These qualities seem theoretically easy to obtain in a material so tractable as soft pine wood, of which drawing boards are generally made. Practically this is very difficult, as wood, however well seasoned, is continually changing its form, rapidly absorbing moisture from the atmosphere, causing expansion of the fibre, and slowly contracting unequally as the moisture evaporates, and this with a force no simple means will resist. For these reasons the true principle of making a drawing board is that which will leave the wood free, so as to allow these changes to take place without materially affecting the surface or square of the board. This is nearly effected in a drawing board invented by the late elder Mr. Brunel. The front surface is quite plain. The construction is as follows :—The board is glued up to the required width, with the heart side of each piece of wood to the surface. A pair of dry hardwood ledges are screwed to the back side. These screws pass through the ledges in oblong slots, bushed with brass, which fit closely under the heads, and yet allow the screws to move freely when drawn by the contraction of the board. To give the ledges power to resist the tendency of the surface to warp, a series of grooves are sunk in, half the thickness of the board over the entire back. These grooves take the transverse strength out of the wood, and allow it to be controlled by the ledges, leaving at the same time the longitudinal strength of the wood nearly unimpaired."

Oak and teak are undoubtedly the best woods for a small plane surface, but provide too hard a table for fastening drawing pins. Deal and other soft woods imbibe moisture quickly, and expand across the grain, so that the protection provided by the best made drawing boards is essential. With regard to the use of the ordinary instruments employed for plotting in office drawing generally, the most advisable course for a student to adopt is to purchase a new case of instruments from some well-known reliable makers, such as Messrs. Elliott Bros., or certain other makers, and he will find with a little practice, observing the essential rule for geometrical drawing, viz., to keep the drawing pen at right angles to the paper, that he will soon begin to ascertain their several uses and value.

Facility in the construction of geometrical figures is given to a draughtsman by the use of what are known as Marquois scales, which are described and illustrated in Heather's Treatise upon Mathematical Instruments, 14th edition.

In the treatise upon mathematical drawing instruments by Mr. W. F. Stanley, will be found a description of various other instruments for producing ellipses, parabolas, hyperbolas, spiral lines, and different ornamental as well as geometrical figures. These instruments are exceedingly ingenious, and a study of their principles will suggest many useful hints to the student, but to possess them would not only occupy more space in an office than can usually be allotted for such storage, but would need the expenditure of more capital than their employment would justify. (See page 110.)

For extensive plans, machine-made papers should be avoided, as they distort under the influence of variation of temperature. Surveys may be plotted upon the best handmade rough double elephant size drawing paper. The paper used for original plans should be mounted on cloth before the plotting is commenced, otherwise the mounting will alter its dimensions, and it should be well seasoned. Drawing paper mounted (hot pressed) upon brown holland can be obtained in a continuous roll to any size if ordered from a good map mounter or stationer. When once laid down to plot a plan, the portion cut off the roll for drawing upon should preferably not be again rolled up, but kept flat and covered over each evening after work, until the whole survey is completed and plotted. Mounted paper is thus found to be less affected by changes of temperature than that which has not been damped, and the effect of being hot pressed is to render the surface smooth, so that fine lines may be drawn thereon.

For small plans showing any enlarged details a sheet of

Whatman's paper may be employed, mounted upon a drawing board in the following manner: lay the sheet of paper upon the board first, with the face uppermost, which is to be finally laid upon the board, and, therefore, the reverse of the surface which is to receive the pencil lines. Take care to observe that the surface of the board as well as its edges are perfectly clean. If the remnants of a previously mounted sheet remain upon the edges of the board, lay a damp cloth over the pieces to be removed, and pass an ordinary laundress's flat iron, moderately heated, over the damp cloth, which will be found to sufficiently melt the glue to enable the paper thus attached to the board to be easily removed with a knife. Precaution should be taken that in so doing the board is not cut by the knife in applying its edge to remove the glued pieces. Afterwards wash the edges of the board with hot water, and when they are dry, lay the sheet of paper upon the board in the manner described above. Then damp the paper equally over by passing a wet clean sponge, not too full of water, but amply moist, first upon the edges about two inches all round, and then over the whole interior surface, after which, leave the paper in this condition for a little while, and look after the state of your glue. Add water as you will have found it has thickened by evaporation since the last operation, and it is necessary to apply the glue to the paper thin, hot and clean. By this time the paper will have soaked in its applied moisture and can then be turned over and set in its position upon the board. The edges of the paper are next turned up against a wood straight edge for about three-eighths of an inch and the liquid glue applied to the turned-up edges. These edges are then pressed by the fingers upon the board all round the paper, care being taken during the operation to stretch the paper as much as possible by pulling the glued edge so that it slides as much as required towards the edge of the drawing board. Attachment to the board may be assisted by rolling the smooth broad edge of the china cover to a cabinet nest of saucers of large size over the edge that is glued. Then finally damp the upper surface of the paper with the moist sponge all over, except the glued edge, and leave the mounted paper to dry, when the contraction (if care has been taken to stretch the paper in mounting equally and sufficiently) should leave the surface quite flat and smooth. If after the paper lies perfectly smooth, the paper in the process of plotting appears to buckle up, owing to a damp atmosphere or variations of temperature, the board should be allowed to stand with its paper face opposite and near, but not close, to a fire, when it will again contract and lie smooth. After use, a draughtsman's gluepot should be covered up, to keep the glue free from dust. Fresh glue needs to be soaked in cold water for about eight hours previous to use. For large plans the paper upon which a survey is to be plotted should never be strained on a drawing board, because when the drawing is cut off the board, and so relieved from strain, the paper will contract, and the plan will, perhaps, contract unequally in all directions. For rough purposes, such as trying the accuracy of base lines at an hotel in the country, or before plotting them upon the best paper, continuous cartridge paper is generally employed. Cartridge paper will not however bear the application of ink eraser or the use of pocket-knife to correct mistakes, and will not take colours or tints nicely; but for the above preliminary work, these excellencies are not required.

Parchment and vellum are more durable than paper; hence we find them employed for old plans where the estates were supposed to be handed down to posterity with their main features unaltered. Vellum exceeds parchment in durability, and it may be necessary to remark that when either of these materials is used for copies of plans, the surface to be drawn upon must first be rubbed with clean flannel dipped in the best Paris whiting. This operation clears the surface from grease, and facilitates thereby the movements of the pen.

In copying plans upon vellum or parchment, place the plan upon a sheet of glass with the parchment over it, so arranged as only to admit light through the glass underneath, shutting out the light above by placing a board in front and also at the sides. A piece of thick clear glass should be used, inclined at an angle in the same way as a reading or music desk. Then every line will appear as distinct as through a sheet of tracing paper. Tracings for the patent office are made on parchment in this way, on paper $29\frac{1}{2}'' \times 21\frac{1}{2}''$, without tints or washes, these being the outside dimensions, so as to permit of $1\frac{1}{2}''$ margin outside a border line $26\frac{1}{2}'' \times 18\frac{1}{2}''$. Plans of greater extent cannot be conveniently copied by this method.

The experienced surveyor will always decide to try his lines upon paper as the work progresses, so that in the event of any error in the lines of construction being committed in the field, the circumstances attending the measurement will be fresh in his memory, and he will be enabled to rectify it with less trouble than if delayed. In an extensive survey the base lines measured each day should be laid down and proved at night, that any error which may have occurred may be at once corrected before the work proceeds further. Some might, perhaps, at first sight deem this tedious and superfluous, but the satisfaction given to a surveyor when his lines meet correctly fully repays for the additional labour.

The best form of beam and compasses for plotting long base lines from chain measurements only, is shown in fig. 12, pages 99-101, the top table forming the letter T, giving lateral rigidity to the beam, to avoid buckling. When only a straight edge is used to hold the beam compasses for a long measurement, it is apt to become bent horizontally while striking in the arcs, and the length of the radius is thereby shortened. To prevent the clamping-screw making an impression on the straight edge, which is used as a radius, it is constructed to work against a plate guided by two steady pins. Fig. 13 shows two other forms of beams, one of a triangular shape, and the other of a tubular shape, the latter with a universal centre-piece marked L, the top of which fixes into one of the boxes in place of the points E or F. The tube between M and N slides within the tube between K and M, and the tube holding B slides within the centre tube from M to N. Clamping-screws marked K, M, N, are provided for fixing the tubes in position at the required distance. (See pages 99-101).

Tubular beam compasses possess the advantage of

portability, and though steadier in their action than when the ordinary beam compasses (fig. 12) are employed, working upon a thin straight edge of flat section, they are not so steady as the ordinary beam compasses attached to a T-shaped beam. Moreover, the stand upon which the instrument is mounted, by concealing the centre from which the circle is to be drawn, is likely to lead to inaccuracy, and is not therefore so satisfactory as when a plain horn centre is employed for the fixed point of the beam compass to turn upon.

In the illustration "Plotting a Plan" (pages 92-94) the plotting scale appears placed at about 80 links from the station A, upon the line A C, in order that the intersection of the arcs at A when struck from the stations B and C respectively, may appear clear in the diagram. The arcs are drawn with the pencil leg of the beam compass and their intersection marked with a needle pointer. When inking in a base line it must be remembered that although it is impossible to represent any measure, which mathematically fulfils Euclid's definition of a line, as that which has length without breadth, yet we should practically so regard a baseline, and hence should draw the line upon the paper as fine as possible, the steel drawing pen being held upright. Base lines intersect at stations, which are each represented by a point to indicate position only, and this should be so clearly defined as to be practically without length, breadth or thickness, and determinable only by means of measurement from other station points as shown in the diagram.

It is impossible to make a correct plan if any of the base lines are plotted too thick or in any way crooked. Hence the long flat rulers intended solely for this purpose are termed "straight edges," and their accuracy should be tested before use. This may be effected by first ruling a line, and then reversing the straight edge end for end, in order to see if the same edge coincides with the pencil line previously drawn, or by placing two straight edges close together and observing if their edges coincide both before and after one of them is reversed end for end when placed in contact. The author uses a steel straight edge, and has in his office a long vertical box, 7 ft. × 1 ft., lined with

PLOTTING A PLAN.   108

baize, a coarse kind of open cloth, to keep the steel free from rust. This box is screwed to the side of an office bookcase, and is air-tight. Steel straight edges are more to be relied upon than wooden ones, even when an ebony edge is attached to the latter. In plotting, a flat ivory or boxwood scale is laid upon the paper exactly parallel (see fig. 1) to the base line, and is kept in position by flat weights resting upon its outer edge. A short rectangular flat scale rests upon its inner edge from which to measure the offsets. These offsets are marked on the paper by pricking with a needle point (see fig. 14, p. 101) opposite the proper graduations on one of its edges, the chain distances at which they are measured on the base line being taken off the long scale. Care must be taken that the edges of the offset scale are truly perpendicular to each other, and that the zero on the offset scale slides exactly over the base line. The advantage of using a split offset scale, or one with zero in the centre, is explained in fig. 1. Metal weights require covering, either of baize or leather, otherwise they cover the drawing with marks which cannot easily be removed (see pages 92-94).

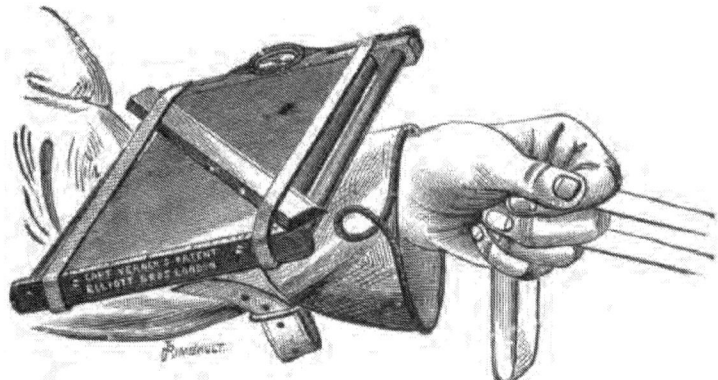

Verner's Wrist Plane Table (or Cavalry Field Sketching Board)—7½ in. × 5 in., with patent rollers, and strap to fasten on wrist, scale of 1 in., 4 in., and 6 in. = 1 mile in yards; normal scale of H. E. for 20 ft. at 6 in. = 1 mile; self-clamping Clinometer and 1-in. bar needle Compass, with direction line, and divided each 10 deg. on ring; complete with ruler and rubber bands, is a useful help in military surveying.

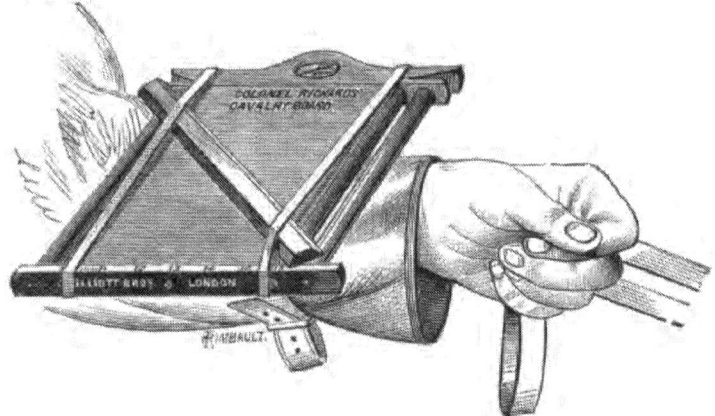

Enclosures are frequently covered with parallel lines in light ink by a process termed hatching. To assist in drawing these lines, an ingenious arrangement suggested by Harden, consisting of a set square and straight edge, has been employed, connected together in a special manner. The distance is adjusted by stops, one formed by the stud on the scale plate fitting the slot in the set square, the other is a projection on the straight edge, the distance between the two being regulated by the scale plate. The straight edge and set square are moved alternately, each guided by the other, drawing a line at each alternate movement. The practical draughtsman would, however, prefer to estimate his own distances, ruling in the parallel section lines off the edge of one set square which slides along the edge of another set square, the latter being maintained firmly in a set position. In the application of Harden's method to cylindrical shading the straight edge has a projecting arm working between a stationary stud and an eccentric stop as part of the set square. The eccentric and scale govern the distance between the lines drawn, limiting the motion of the arm between the stops. The eccentric is revolved equal distances according to the scale of degrees, this giving an increasing distance between the lines drawn.

Corey and Barczinsky's Technical Drawing Apparatus consists of a T square (42 inch or 32 inch) with two stocks, one of which is fixed, and the other movable. The blade of the T square has two inverted T grooves, in which slides the dial. This dial which is made of brass is graduated into half-degrees, and is fitted with an indicator and arm, or ruler, also of brass (these latter in one piece, and so arranged that when the indicator shows a certain angle, the ruler forms that angle with a perpendicular to the edge of the T square), at the important angles, viz.: $30°$, $45°$, $60°$, $67\frac{1}{2}°$ and $90°$. There are special divisions near the edge into which a small knife fitted to the indicator drops and fixes it firmly, the degrees being read through an aperture in the end of the indicator. A small screw lifts the knife out of the division when the angle is wanted to be altered This instrument is a combination of T square, set square of all angles, and protractor; and besides doing the work of each and all of these, it will divide circles or parts of circles into any number of equal parts without calculation. The instrument can be applied to the drawing board by loosening the movable stock, pressing it tightly to the edges of the board and screwing up, when it will travel easily up and down, and thus all horizontal lines are made. The dial slides from side to side in its grooves, and therefore by these two movements it may be placed in any position on the surface of the board, consequently lines either vertical, horizontal, or at any angle, may be drawn. To use the instrument as a set square of any of the common angles such as $30°$, $45°$, $60°$, $67\frac{1}{2}°$ and $90°$, the knife is raised out of the divisions on the edge by the small screw, and the indicator is moved over the dial until the angle required is seen through the aperture; then by releasing the screw, the indicator will be fixed at that degree.

Such instruments display skill in their construction, but are seldom near to hand in an office when required as they are seldom used, and the remarks made upon page 103 apply thereto.

CHAPTER XI.

COMPLETING AN ESTATE PLAN.

THE accompanying specimens of typographical hieroglyphics may serve to guide the student as regards conventional signs. (See pages 114-119.) The writing upon a plan should be disposed in parallel directions reading from west to east, or south to north, with the exception of the names of rivers, canals, chains of mountains, etc., which should be adapted to their natural sinuosities. Dotting pens, or pens fitted with wheels for drawing in lines representing boundaries that may be walked over, do not work sufficiently regular in their action to produce neat work. With practice, the draughtsman can speedily draw a dotted line neatly.

Major E. R. James, R.E., in his remarks upon hill sketching, published in "Ordnance Maps: the Methods and Processes Adopted for their Production," states that, "to give pictorial effect and softness of expression to a sketch, the penmanship requires skill in various minor points, which can be acquired by practice only. These are the graduation in the thickness of the stroke in the expression of slopes of gradually varying altitude; the evenness or raggedness of the edges of the strokes; and the manner of disposing the breaks between the ends of the touches. The latter point especially requires great attention, and the breaks between strokes should never form continuous white lines, which attract the eye. Again, if the strokes be long and even and the curves gradual, the ground represented is of a smooth surface; if the contrary, of a rough surface. Rocky features being expressed by vertical strokes, the roughness of appearance is materially aided by their introduction. The light is supposed to fall vertically, that is, slopes of like altitude and inclination are similarly expressed, whatever may be their compass bearings."

In order that colour may flow easily, and cover a surface evenly, it is necessary that it should be thin for tinting, as the draughtsman must remember he has not to paint. It is always easy to wash it over again if it is not dark enough, but it is very difficult to wash off the colour if it be too dark.

Wherever it is possible, use a large brush in preference to a smaller one, as you will by this means be more likely to succeed in getting a flat wash, whilst a small brush might make the tint lie in streaks, as it will not complete the surface to be tinted with sufficient rapidity to obtain a flat wash. Care is, however, necessary in using a large brush, so that you may not pass over the outlines. Take care to prepare an ample quantity of the correct tint, and when you have rubbed as much colour as you think you are likely to want, do not at once put the cake back into its place in the box, but stand it on one of its edges so as to allow it to dry, otherwise it will stick to the box.

Blue, red, and yellow are called the three primary colours. When two primaries are mixed they produce a secondary colour. Thus:—

Primaries.		Secondary.
Yellow and Red	produce	Orange.
Yellow and Blue	,,	Green.
Red and Blue	,,	Purple.

When you wish to mix a secondary colour, such as green, from the two primaries blue and yellow, rub the blue in one division of the slab, and the yellow in another, leaving a space between them. Then, with your brush, mix the two colours in this vacant space; but on no account rub either of the cakes in the colour obtained from the other, as this would leave the end soaked in another tint, and when you use it again you would find the colour would be impure. Of course, these remarks apply to the mixing of any two colours.

The following is a conventional list of the colours used by most surveyors to express the various substances usually to be dealt with in the completion of the Plan of an Estate:—

Material.	Colour.
Brickwork to be executed (in the plans and sections)	Crimson Lake.
Brickwork in Elevations	Crimson Lake mixed with Burnt Sienna or Venetian Red.
The lighter Woods—such as Fir	Raw Sienna.
Oak or Teak	Vandyke Brown.
Granite	Pale Indian Ink.
Stone generally	Yellow Ochre, or Pale Sepia.
Concrete Works	Sepia with dark markings.
Wrought Iron	Indigo.
Cast Iron	Payne's Grey or Neutral Tint.
Steel	Pale Indigo tinged with Lake.
Brass	Gamboge.
Lead	Pale Indian Ink tinged with Indigo.
Clay or Earth	Burnt Umber.
Slate	Indigo and Lake.

To mix Indian ink, place a small quantity of clean water in a clean saucer, rub the ink round, holding the stick vertical, and as soon as the bottom of the saucer is made visible by the stick of ink passing over it, add a little more water and stir round.

When inking in a plan, it will be found that breathing between the nibs of a clean pen before immersing it in the ink is preferable to the employment of a camel hair brush *veying and Levelling, pp.* 114, 115, 116.

	MEADOW & PASTURE LAND
Ma	*May be coloured light green.*
	COPPICE WOODS
C	ORCHARDS

Surveying and Levelling, pp. 117, 118, 119.

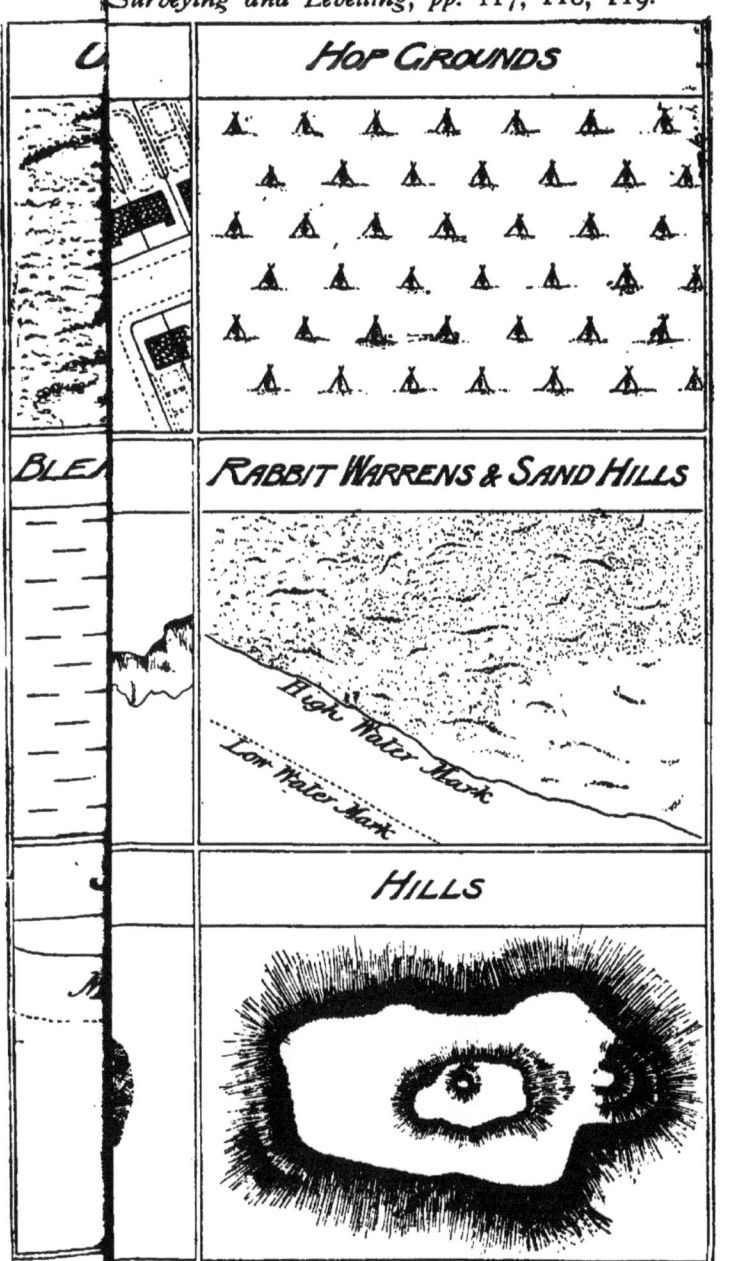

for filling the pen. Care must be taken to wipe the outside edges of the pen, so as to prevent the ink flowing upon the edge of the instrument against which the lines are drawn. A folded piece of clean blotting paper is the best for this purpose, and better than an office duster, which in a forgetful moment might transfer the moist ink to the drawing paper. The blotting paper absorbs the ink quicker than a cloth, and hairs are apt to get into the ink of the pen, as is often the case when a fluffy cloth is used. A feather-edged scale should not be employed as a ruler, as the edge is not thick enough to prevent the ink escaping upon it, and so smudging the drawing, unless the scale is turned down upon its face with its broad side uppermost. Common ink ought never to be used in planning, because it sinks too deep into the paper, and in process of time becomes discoloured.

To finish a drawing with Indian ink alone, so as to delineate all the outlines and deep shading, needs good draughtsmanship and takes time. Some surveyors prefer to make free use of various tints, employing a light green for pasture, meadows, bogs, morasses, trees, &c., with Prussian blue for giving dark shades to the sides of rivers and the sea shore, a little yellow mixed with a little lake colour for shading roads, and tints of umber and burnt sienna for different shades of brown. Weak Indian ink answers for the first tints to delineate hills, taking care to wash the top and bottom off with clear water and a clean brush, to soften the edge before it has time to dry. A little Prussian blue, judiciously mixed with the Indian ink, contributes a good effect. If the hills are very steep, add another tint or two, and shade them according to the steepness, never neglecting to soften off the edges while wet with a clean brush and clear water, and always allowing each coat to dry before another is applied. Rocks may be tinted resembling the colour of the stone by shades, and made to appear rugged. To indicate morasses, use a fine pointed camel hair brush with light Indian ink, and draw with the hand short horizontal lines, fairly close to one another, some long and some short. To represent gravel, mix light Indian ink applied to a tooth brush, and by the aid of the edge of an old scale, slightly touch the hairs and thus spirt it over the part of the plan required, having first cut out a piece of tracing paper the form of the area to be delineated, in order to cover the adjacent parts of the plan and protect them from becoming marked. Rushes and reeds, &c., can then be inserted with a pen and the whole shaded with a pale green, tending to a blue tint. Finally touch up with a strong green and shade off towards the right hand of the plan with lighter green. Meadows are shown with long and short dark strokes and washed over with light green tending to a yellow tint. Pastures have shorter dark strokes, some sloping, some upright, and not exceeding the 50th part of an inch, run over with horizontal shades of Indian ink and washed when dry with a green tint somewhat darker than the meadows. Houses, slept in, are tinted with carmine and lake, outhouses with moderately light Indian ink, except glass enclosures, which are tinted a light blue. For trees and hedges we may employ Indian ink and yellow and indigo. Trees are indicated by a vertical stem, with a horizontal shade at the base, made widened out at the top, and shaded either with the pen upon the right or east side or with a touch of light Indian ink and tinted green on the left or west side. If the trees are numerous, variety may be introduced by touching up the west side of some of the trees with a little brown, others with a little yellow, instead of green universally. Ploughed lands are represented by drawing narrow parallel lines representing the furrows. Cornfields need to be done neatly to produce a good appearance. Parallel lines at approximately equal distances are drawn with a drawing pen by hand, in weak Indian ink or with the colour selected to shade the field, for representing the ridges. Then with an approved colour, say yellow, run down the ridges, softening off the edge as before with clear water and a clean brush. In the case of adjoining fields the tints may be varied between light brown and yellow. Sands may be shown with a series of small dots made with a writing pen, and when dry, first washed over with clear water and immediately dried with a clean sheet of blotting paper to remove the surplus ink, and then tinted over with a little weak carmine and gamboge or a coat of light burnt sienna. Rivers, lakes, and the sea shore are shaded by tinting round the edges on both sides with fairly strong liquid Prussian blue, which is softened off with a hair pencil, which should never be put in the mouth, but moistened with clean water and passed over blotting paper by twisting round thereon, in order to obtain a fine point and to remove surplus moisture. Broader belts of lighter blue are applied towards the middle of the river. (See page 138.) Gardens are tinted brighter or darker green than fields or grass. (See pages 114-119.)

Proficiency in tinting plans can only be acquired by practice and by observation of an experienced draughtsman at work. Great care must be taken in delineating the various ravines and brows of hills, and also not to labour the hills too much, as the best effects are obtained by a few successful touches, as it must be carefully borne in mind that a surveyor's plan should never be coloured, but tinted. In making the enlarged plan of a building, the name of every room, office, yard, etc., should be stated either within the rooms themselves, or referred to in the margin of the plan by letters. The doorways may be shown by leaving the space they occupy in a wall untinted or not hatched by oblique lines, the window sills represented in the same way, the chimney bottoms or fireplaces denoted by making the inside of the wall to project into the room at right angles, and the steps of the staircase exhibited by parallel lines, or radial lines for a winding staircase, in either case drawn at proper distances apart and numbered from the lower landing or floor. The inside of the rooms may be left white or tinted at the option of the draughtsman.

Land surveying implying the measurement of everything occupying the surface of the ground, it is clearly inconsistent to show an archway upon a plan as a block building, notwithstanding that the building above the arch may form part of the same house as exists upon one or both sides of the archway, and inasmuch as it is incorrect to show superstructures upon a map or ground plan which do not actually rest or bear on the ground, it is equally wrong to show any underground work by a firm line. Underground works and portions of superstructures which have free openings between them and the ground, should be delineated on the plan by finely dotted lines. Also the edge of a footpath is shown dotted because there is no obstruction in such boundary. It can be walked over.

It is recommended to add the following note to a building plan of a new estate, thus:—

Building Plan
of
....

The property of the.......................................

.... Designed by......................

NOTE.—This plan is designed to show the general position of the sites of the houses and villa plots, and is subject to such variation as may be found desirable.

The Surveyor must take care that the north point and the scale of the Plan are both neatly and correctly drawn in suitable positions upon the Plan. If a chain of 66 feet (100 links) has been employed for the survey, it is well to add an equivalent scale in feet, as shown in figure 1, pages 93, 94. The scale adopted for the Ordnance survey are indicated (full page) upon pages 124-126.

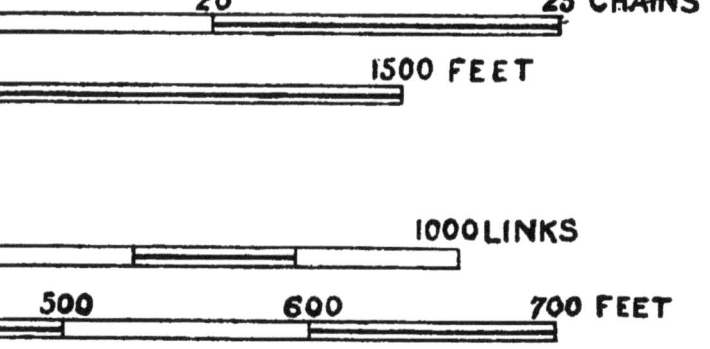

Land Surveying and Levelling, pp. 124, 125, 126.

CHAPTER XII.

ORDNANCE MAPS

A TRIGONOMETRICAL survey of the United Kingdom was commenced in the year 1784, by General Roy, in the measurement of a base line upon Hounslow Heath. The base was intended at the time mainly for astronomical purposes, and was measured thrice—viz. (1) by cased glass tubing, (2) by thoroughly seasoned deal rods in 20 ft. lengths, cut out of an old Riga mast and trussed, (3) by a 100 ft. coffered steel chain, specially made by Ramsden. The three measurements tallied with the result of subsequent re-measurement very closely. While the line was over five miles in length, the greatest difference in any of the admeasurements in 1791 was under six inches. It was in that year (1791) that the Government determined upon the construction and publication of a military map of the United Kingdom, on the scale of one inch to a mile, but it was not until 1797 that the design of a general map founded on a minute survey was first conceived. In 1794 a base line of verification nearly seven miles long had been measured upon Salisbury Plain with a steel chain, and in 1798 another base line of verification was measured at Sedgemoor, in Somersetshire, a trifle under $5\frac{1}{4}$ miles long, the triangulation being meanwhile extended throughout the country traversed, by means of which about 250 main points, situated in various places, were trigonometrically determined with great accuracy. By 1824 the whole south of England and parts of Wales and Scotland had been surveyed. At the time of the death of Colonel Mudge, under whose direction the first sheet had been published in 1801, it became necessary to appoint a successor. This occurred in 1820, when Captain, afterwards Major-General Colby, who had hitherto assisted, was then appointed to direct the survey, and from that year (1820) gave his whole energies to geodesy and topography. He introduced compensation bars, 10 ft. long, formed of iron and brass, combined in such a manner that the distance between two fine dots, upon steel tongues connecting them at their extremities, was calculated to remain constant at all temperatures. In connection with two principal base lines of the triangulation, situated, one at Salisbury Plain, and the other on the border of Lough Foyle, the bars were laid perfectly horizontal, and in a perfectly straight line, by the aid of levels and a directing telescope. In order to avoid the disturbance of a bar when once in position, the next, instead of being brought into contact with it, was placed so as to leave a space of six inches between their ends, this interval being measured with a double microscope, the two foci of which were kept at all times exactly six inches apart by means of a compensating apparatus constructed upon a similar principle to the bars themselves. Colby directed the survey until his retirement in 1846, and both before and after this date, its progress has mainly depended on the sums voted by Parliament for its execution. Thus in 1824, when the Government decided that the general valuation of the land of Ireland should be based upon an accurate map, the survey in Great Britain became practically suspended. For the purpose of valuation the scale of an inch to a mile was obviously much too small, and a ground plan to a scale of six inches to one mile was therefore adopted, which was almost completed in 1840, when it was decided to adopt the same scale for the unsurveyed parts of Great Britain. The one inch map was not abandoned, but instead of being obtained by a special survey, it was now produced by reduction from the six inch sheets. The addition of contours to the six inch map provided accurate data upon which to base the shaded delineation of hills upon the one inch map. Horizontal lines, or rather lines of equal level, were actually traced with the levelling instruments at vertical intervals of 100 ft., subsidiary contour lines at vertical intervals of 25 feet being sketched by the surveyor on the ground. The scale of 3 chains (198 feet) to 1 inch, upon which most of the tithe maps were drawn, or $\frac{1}{2376}$, proved itself more useful, where applicable for the various kinds of local administration, than the scale of 6 inches to 1 mile, or $\frac{1}{10560}$, and this led vendors and purchasers and others interested in the uses of valuation for public and private purposes to agitate in favour of a scale not too minute to be employed in the conveyance at least of agricultural land. The battle of the scales culminated in the International Congress, held at Brussels in 1853, when the scale of $\frac{1}{2500}$ was agreed to be adopted for cadastral* purposes, that having been the scale sanctioned by Napoleon I. for the cadastral survey of France, and recommending itself to Britishers as practically giving one square inch to the acre. As a matter of fact, a square of 1·0018 inches in the side contains an acre, but when the shrinkage of the map is considered the above-mentioned ratio of an acre to the square inch of paper is approximately correct. This decision involved a new survey to a scale of 25·344 inches to a mile for the cultivated districts, and the sheets on this scale were henceforth used for the construction of both the smaller scales. This change also resulted in omitting contour lines from the (so-called) 25 inch map and adding them at 100 feet intervals to the map reduced therefrom to form the 6 inch map. For the triangular connection between England and the Continent, a repeating circle of Borda's make, about 12 inches in diameter, was used on the French coast, and a Ramsden 3 feet theodolite upon the south coast of England.

The designation Ordnance Survey is due to the fact that at first the directors of the survey were responsible to the Honourable Board of Ordnance, the control passing in 1855 to the War Office, and in 1870 to H. M. Office of Works. The Ordnance Survey is now a branch of the Board of Agriculture.

The various changes made in the work of the survey down to the present time have resulted in a programme which embraces the production of five sets of maps. Plans on a scale of 10·56 feet (126·72 inches) to the mile, or

---

* "Cadastral" (French, "cadrer" to square or correspond with; Latin, "caput" a head) implies the application of assessment or revenue to the areas computed from the survey to which this designation is applied.

$\frac{1}{500}$ (41·66 feet to the inch, or 1·584 inch to one chain) of all towns with more than 4,000 inhabitants. These are sufficiently large to indicate detail down to the size of a doorstep and an area grating. This scale was determined upon in 1855 as being lineally five times that selected for cadastral purposes, but for London and its environs five feet or 60 inches to the mile was retained ($\frac{1}{1056}$). Each town is provided with an index or key map, so that the town sheets may be readily selected in purchasing maps. The illustration on pages 124-126 shows the Ordnance map scales, furnished with a scale of chains in links. In scaling distances upon any printed map it is well to use the actual scale drawn on that map. For Ordnance maps, shrunk scales can be obtained, which are made either in boxwood or ivory. The shrinkage is stated by Mr. W. F. Stanley, the instrument maker, to be about one-seventh of an inch to the foot run. The junctions between Ordnance sheets follow the cardinal points of the compass, and the distances between the parallel lines north and south, east and west, for the sake of comparison, are indicated upon a diagram given on pages 132, 133, which show the manner in which Ordnance maps are numbered. Parish maps on the scale of 25·344 inches to the mile are arranged by counties. The quantity of ground shown in a parish sheet is the sixteenth part of the quantity of ground included in a county map, or one-sixteenth of twenty-four square miles; that is $1\frac{1}{2}$ square miles or 960 acres. Indexes to the plans of the parishes are bound up in the area books, by which the number of the sheets, including any particular locality, may be seen, Each enclosure is distinguished by a number referring to a tabular index, which gives its description and its acreage in the area book.

County sheets give convenient sized maps of 3 feet by 2 feet representing 6 miles by 4 miles, and containing 15,360 acres. Indexes are published on scales of 2, 3, or 4 miles to the inch, and from these the relative positions of the 1 inch, the 6 inch maps, and of such quarter sheets of the 6 inch maps as are published, together with the sheets of $\frac{1}{2500}$ scale may be identified. (See diagram, pages 132, 133.)

The maps to the scale of 1 inch to a statute mile form

a general map to a scale of $\frac{1}{63360}$, useful for topographical purposes. No fences are shown on this map. The width of the roads and size of the houses are greatly exaggerated for the sake of clearness. Four kinds of roads are indicated, (1) metalled roads between towns, (2) roads metalled and roads in good repair between a town and a village, also between a village and a village, as well as important cross roads from one town road to another, (3) minor roads, blind roads, carriage drives through ornamental grounds, and well-made roads in good repair to important farms, (4) bridle roads, being not mere cart tracks, are marked with the initials B.R. The maps present a correct picture, exhibiting contemporary distribution and general direction of boundaries.

There are two series, known as the old series and the new series. All the counties in England and Wales, except the four northern counties and the northern halves of Yorkshire and Lancashire, have been published in outline with the hills shown by vertical hachures and sold in sheets and quarter sheets. Sheets of this series coloured geologically are also issued. Indexes can be obtained to scales of 10–30 and 50 miles to one inch, showing the number and position of the sheets, but the large scale of 10 miles to the inch is the most useful, as it also furnishes the name of every parish and town. The sheets of the new series are published either hill shaded or with the form of the ground delineated from contour lines. The details shown on these maps are the same in both cases. The size of the sheets is 18 in. × 12 in., representing an area of 216 square miles. (See pages 132, 133.)

The parish maps being reduced by photography for the foundation of a county map, the one inch maps are mainly reduced therefrom by the pentagraph. The town map of $\frac{1}{500}$ is similarly reduced to $\frac{1}{2500}$ for representing the towns upon parish maps previous to transfer to zinc plates for publication. The county maps on the six inch scale down to the year 1882 were all engraved on copper, but since that date they have been mostly published by photo-zincography. The process of photo-zincography combines the accuracy of photography

ORDNANCE MAPS.

DIAGRAM No 1. — A 6" Map showing 5 Feet Maps.

1	2	3	4	5	6	7	8	9	10
11	12	13	14	15	16	17	18	19	20
21	22	23	24	25	26	27	28	29	30
31	32	33	34	35	36	37	38	39	40
41	42	43	44	45	46	47	48	49	50
51	52	53	54	55	56	57	58	59	60
61	62	63	64	65	66	67	68	69	70
71	72	73	74	75	76	77	78	79	80
81	82	83	84	85	86	87	88	89	90
91	92	93	94	95	96	97	98	99	100

VII (central)

↕ Four Miles ↕ — Actual Width represented upon a 6" Map — Six Miles

NOTE — EACH 6 INCH MAP IS NUMBERED IN ROMAN NUMERALS (THUS VII). THE AREA COVERED BY EACH MAP ON THE SCALE OF 6 INCHES TO A MILE IS DRAWN ON ONE HUNDRED SHEETS ON THE SCALE OF 5 FEET TO A MILE — AND THE 5 FEET MAPS WITHIN EACH 6 INCH SHEET ARE NUMBERED IN ARABIC NUMERALS (THUS 53) FROM 1 TO 100. IN ORDER TO IDENTIFY ANY 5 FEET SHEET THE NUMBER OF THE 5 FEET SHEET AND THE NUMBER OF THE 6 INCH MAP IN WHICH THE 5 FEET SHEET IS SITUATED MUST BE STATED (THUS VII.53). A 5 FEET = 1 INCH SHEET IS THEREFORE ⅗ MILE (3168 FT) × ⅖ MILE (2112 FT) IN OUT-SIDE MEASUREMENT IN THIS INDEX THE CENTRAL NUMBER VII REFERS TO THE 6 INCH MAP AS EXPLAINED ABOVE. —

THE ROMAN NUMERALS IN THE 16 RECTANGULAR SPACES (XXXVIII) REFER TO THE 25 INCH MAP & THE ARABIC FIGURES IN THE LOWER DIAGRAM SHOW HOW

DIAGRAM No 2. — A 6" Map showing 25" Maps. —

XXXVIII	XXXIX	XL	XLI
XLIX	L	LI	LII

VII

↕ 4 Miles ↕ — Actual Width represented upon a 6" Map

Land Surveying and Levelling, pp. 132, 133.

AGAIN DIVIDED INTO 25 MAPS TO A SCALE OF 1/500 WHERE THEY ARE USUALLY INDICATED IN THE RIGHT HAND TOP CORNER THUS VII 38.21. MEANING MAP № 1 FROM DIAGRAM № 3 OUT OF MAP N°38 (XXXVIII) ON DIAGRAM N°2, OUT OF MAP VII OF DIAGRAM N°1 EACH TOWN MAP, DIAGRAM N°3 IS THEREFORE 1/10 MILE (1584 FT) × 1/5 MILE (1056 FT) IN OUTSIDE MEASUREMENT.—

COUNTY MAPS ARE ORDERED THUS, SHEET VII (DIAGRAM N°1)

PARISH MAPS ARE ORDERED THUS, SHEET VII 38 (DIAGRAM N°2)

TOWN MAPS ARE ORDERED THUS, SHEET VII 21 (DIAGRAM N°1)
SHEET VII 38-21 (DIAGRAM N°3)

LXI	LXII	LXIII	LXIV
LXXV	LXXVI	LXXVII	LXXVIII

— — Six Miles — —
— Actual Width represented upon a 6" Map —

DIAGRAM N°3 — A 25" Map showing 10 feet Maps.

1	2	3	4	5
6	7	8	9	10
11	12	13	14	15
16	17	18	19	20
21	22	23	24	25

— One Mile —
— Actual Width represented upon a 25" Map —
— ½ Mile represented upon a 25" Map —
— Four M —
— Actual Width repre —
— Actual Width represented upon a 25" Map —

with the facility of printing from zinc plates. A lithographic stone is not only more costly than a zinc plate of equal surface, but the former, about $4\frac{1}{2}$ in. thick, could not be lifted about by one man, while the latter, being only about $\frac{3}{10}$ in. thick, can have its surface satisfactorily prepared for the process, and is less liable to fracture. Parish and town maps are therefore now prepared by zincography. A parish map, reduced by photography for the six inch sheets, has subsequently contours added at stated vertical intervals; also the degrees, minutes, and seconds of latitude and longitude are marked upon the margin, by which means the geographical position is determined. The one inch map is reduced from a blue print of a six inch map, in which all that is to appear is inked in with black ink, and is thus taken from a larger scale map and transferred to a copper plate, the hill features, and names, &c., being subsequently added to the one inch map.

A series of one inch maps in colour printed from stone is also in progress of execution, showing the general details in black, the water in blue, hills in brown, contours in red, and the roads in brown. The sheets already published include the southern and south-eastern portion of England. An edition of the one inch map of Scotland, showing the outlines and contours in black and the hills in brown, is also in course of publication by lithography.

It must be remembered that Ordnance maps indicate objects only, and leave their interpretation to owners, and to the judgment of their surveyors. The maps show the fences as they exist upon the ground. They take the centre of the hedge or fence without regard to the boundary of a property. The estate boundary to which the hedge belongs may be 3 ft. or 4 ft., more or less, according to the custom prevailing in the county, upon one side of such hedge. There is no regulation universally applicable thereto. An owner may have a ditch and hedge as wide as he likes, provided he keeps within his own boundary and does not encroach. (See pages 21, 22.)

All the maps are drawn upon one system, the same symbols being uniformly employed throughout, and in order that reference to the maps may be made more easy and

Surveying and Levelling, pp. 135, 136, 137.

SIGNS

C
R
H
P
T

C.S. .... C.S.	1
C.R.	2
S.S.	3
S.D.	4
STATION BELOW S.D. / S.S. / S.S.	5
SIDE of FENCE R.H. R.H.	6
F.W.	7
4 FT R.H.	8
4 FT R.H. 4 FT R.H.	9
OR FENCE ETC 4 FT R.H. 4 FT R.H.	10
4 FT R.H.	11
4 FT R.H. 4 FT R.H.	12
C.F. T.C. C.W.	13
DEF. T.S. UND.	14

AT WHICH A CHANGE OF BOUNDARY TAKES PLACE
STRAIGHT BOUNDARY CHANGES SIDES
STREAM OR DRAIN CHANGES SIDES
BOUNDARY CHANGES SIDES
FENCE CHANGES SIDES
BOUNDARY SIDE OF FENCE
FENCE CHANGE SIDES
DOTS SHOULD BE IN CONTACT WITH THE LINE OF STREAM OR DRAIN, ROOT OF HEDGE, FACE OF WALL ————
DOTS SHOULD NOT BE IN CONTACT WITH THE HEDGE OR FENCE ETC ————
TO BE ON CONTINUOUS LINE REPRESENTING COP, CENTRE OF WALL ETC ————

intelligible, boundaries are marked by clearly defined strokes and dots, and territorial demarcations by characteristic styles of writing for the names, which render them plainly distinguishable. A table of comparative characteristics and symbols in use on the six inch and $\frac{1}{2500}$ scales is given (pages 135-137), but of course upon a plan the thicknesses are drawn less than that shown upon this illustration.

Double lines of railway are shown on the six inch maps by parallel lines with cross strokes. Single lines of railway on the six inch scale are represented by a single line with strokes crossing it; on the $\frac{1}{2500}$ scale both lines are shown.

All first class public roads are shown by the east and south sides from the light being shaded. Occupation or private roads are not shaded. Rivers are marked by the side next the light being shaded, upon the west and north sides. The light is conventionally supposed to enter a plan from the north-west corner. Canals are distinguished from roads by the parallelism of the sides, by the locks and bridges, and by having the side next the light shaded like rivers.

In dealing with the enclosures, braces ⌒ on the plans indicate that the spaces so braced are included under the same reference number. Arrows ⟶ along streams show the direction of the flow of water. When a change occurs in the boundaries, the symbol ⌀ is used to show the point at which the change takes place, as the boundary may change in character.

Sites of antiquities are shown by the symbol ⊹ and to show the periods, the names of such are written thus:—

ROMAN— 𝔓𝔯𝔢𝔥𝔦𝔰𝔱𝔬𝔯𝔦𝔠 *or* 𝔖𝔞𝔵𝔬𝔫 — 𝔑𝔬𝔯𝔪𝔞𝔫 *or Subsequent.*

Altitudes are given in feet above the Approximate Mean Water at Liverpool. Those indicated thus—⊹ *B.M.* 5·47 refer to marks made on buildings, walls, &c., and are called Bench Marks. The use of the very appropriate barbed head or broad arrow ⋏ as the Ordnance mark for levels upon walls and buildings is thought by some to have originated in its having been the crest of a former Master General of the Ordnance (*Notes and Queries*, Dec., 1872).

In a book published by Messrs. Stanford, of Charing Cross, entitled "Ordnance Maps: Methods and Processes Adopted for their Production," the Ordnance datum employed for the operation of levelling is thus described:—
"The datum plane for Great Britain was determined in March, 1844, and is the height of the mean tide at Liverpool, according to the observations made at that time; this height has however since been proved to be 0·068 of a foot above the local mean level by calculations based on the mean of the recorded observations of four successive years, made with the self-registering tidal guage at St. George's Pier, Liverpool.

"The datum plane for Ireland passes through a point, fixed on the 8th April, 1837, on Poolbeg Lighthouse, Dublin Bay, at low water mark of spring tides.

"The adoption of mean water as a datum level for England in preference to low-water mark resulted from a series of tidal observations instituted round the coast of Ireland, in 1842, by the late Major-General Colby, which clearly showed that mean water is more nearly on a uniform level than low-water spring tides, and is consequently better adapted for a plane of reference for the altitudes of a general survey. As from the irregular conformation of the coast line, the local mean level of the sea varies considerably, the general mean for England was obtained by tidal observations made at thirty-two different stations, and for Scotland by observations at eighteen stations, in a similar manner to those previously made at Liverpool. By connecting the mean found at each station with the levelling it was found that the relation of the local to the general mean varies in England from minus 1·283 feet to plus 1·850 feet, and that the general mean level of the sea is 0·650 of a foot above the assumed mean level at Liverpool, which is the Ordnance datum."

Another bench mark employed, which is recognised in the Thames Valley, is known as Trinity high water. This is the level of the lower edge of a stone fixed in the face of the river wall upon the east side of the Hermitage entrance to the London Docks. No permanent mark is at present fixed to denote Ordnance datum, but Trinity high water (T.H.W.) line is taken as 12·48 feet above Ordnance datum. Colonel Sir Henry James, while director of the Ordnance Survey in 1861, published a book, entitled, "Abstract of Levelling in England and Wales," in which he states, "the datum level for Great Britain is the level of mean tide at Liverpool, as determined by our own observations, and it is eight-tenths of an inch above the mean tidal level obtained from the records of the self-registering tide gauge on St. George's Pier, Liverpool." Ordnance datum is thus seen to be an arbitrary level, obtained as above explained, the general mean tidal level of the sea round the coast of England being somewhat higher than the mean sea level at Liverpool assumed by the Ordnance Department. While at Liverpool the mean tidal level of the sea is ·650 feet above Ordnance datum, at Dover we find it to be ·975 above Ordnance datum. The range of the tides is given in a book annually published by J. D. Potter, of the Poultry, London, entitled, "Tide Tables for the British and Irish Ports," and the information therein contained is prospective for the year recorded. Ordinary spring tides and average spring tides are convertible terms. A difference in range of 3 feet found at one place over another shows that high-water ordinary spring tides are 1 foot 6 inches higher, and low-water ordinary spring tides 1 foot 6 inches lower at one place than the other. The tide charts of the English and Bristol Channels and of the entrance of the Thames, by Lieut. A. H. Percy, also published by Mr. Potter, are useful for reference, to show the exact courses of the currents in these channels.

CHAPTER XIII.

ENLARGING AND REDUCING PLANS.

The enlargement of plans by the aid of either the pentagraph or the eidograph can never be recommended where great accuracy is of importance. Where possible, the best way is to replot the whole survey from the field-book to the enlarged scale required. But it may be well, nevertheless, to describe these instruments, as they are of use under some circumstances.

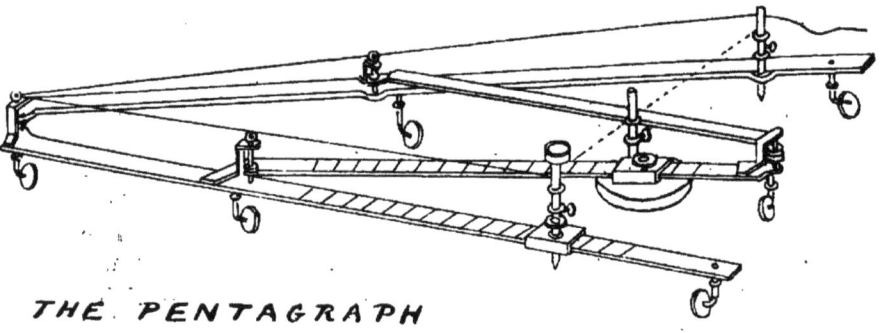

THE PENTAGRAPH

The Pentagraph is usually made of brass, and consists of four flat bars, so fixed as always to form a parallelogram in all positions of the instrument. The instrument works as a jointed rhombus, in which two longer bars shown in the figure are united by a double pivot, which is fixed to the end of one of the bars, and works in two holes placed at the end of the other, forming a knuckle-joint connection. The pencil point is attached to one of these bars, and the tracer point to the other bar. The two shorter bars are fixed by pivots to the longer bars, and are also joined at their opposite ends in a similar manner to the joint uniting the longer bars. Several ivory castors support the machine parallel to the paper and allow it to move freely in all directions. The position of either the pencil or the tracing-point is a fixture, while the position of the fulcrum attached to one of the shorter bars, and the position of either the tracing or the pencil point which is attached to the adjacent longer bar, can be shifted to suit the proportion required for reduction. This adjustment is effected by means of sliding boxes, which can be fixed at any part of the adjacent long and short bars by turning milled-headed clamp screws. The fulcrum contains a lead weight, to which is fixed a bright iron or steel pin, over which the whole instrument travels when in use, being balanced by the six wheels, which

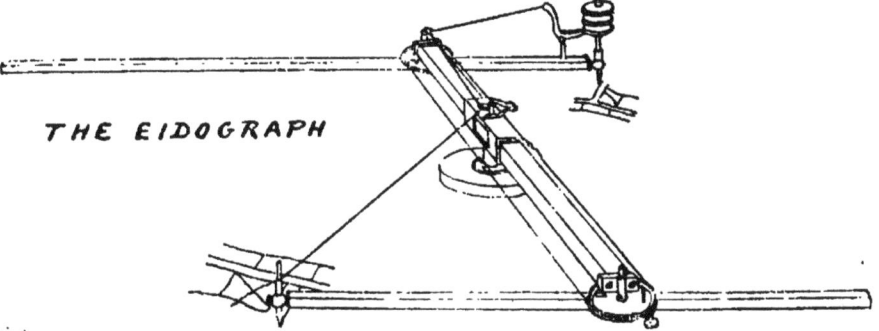

THE EIDOGRAPH

rest upon the drawing paper or table upon which the instrument is working. It is important that the drawing paper should lie perfectly flat upon a level table, otherwise the wheels upon which the instrument is mounted will sustain frequent jerks, which will lead to inaccuracy in draughtsmanship. The principle of its construction is seen by experiment to be due to the fact that two points moving in a plane in any direction, but always remaining in the same right line with a fixed point and preserving the same proportional distance from it, will describe similar lines and figures.

In the Eidograph, invented by Professor Wallace, of Edinburgh, in 1821, consisting mainly of three stout brass bars, these supporting wheels are dispensed with, and the

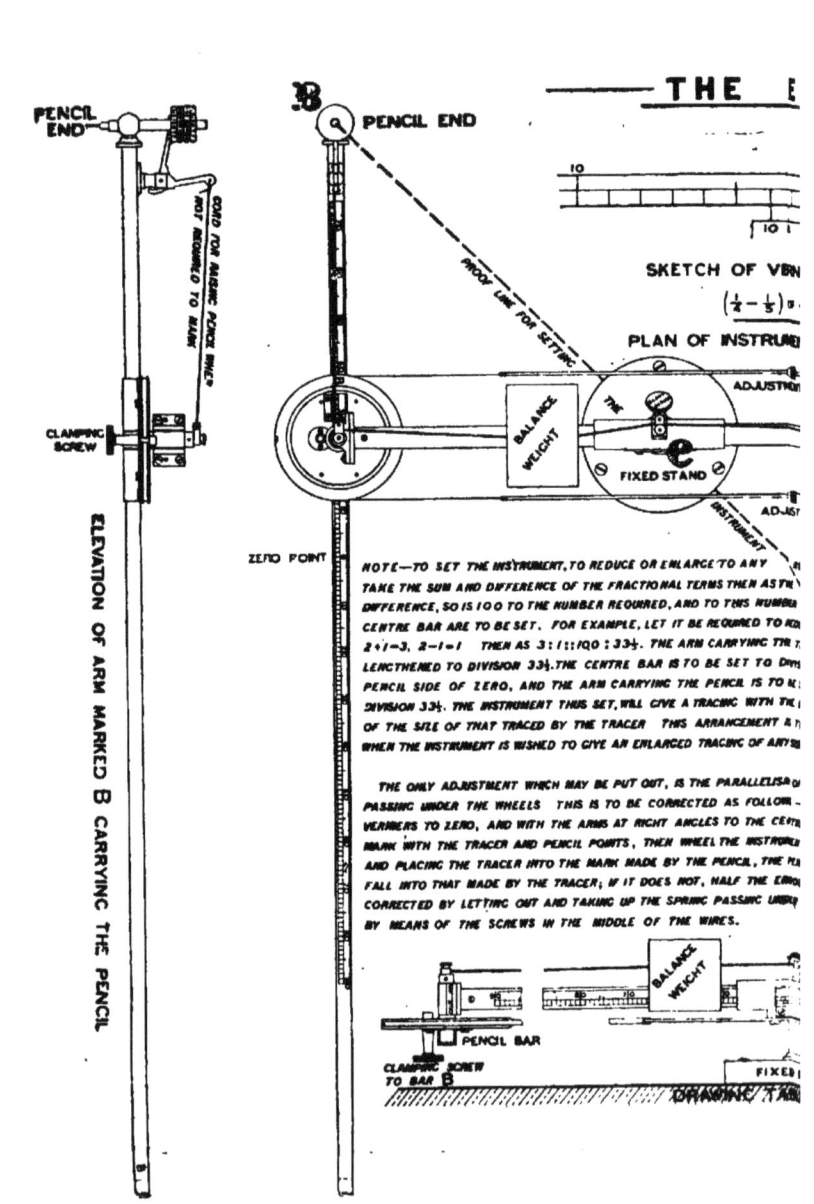

Land Surveying and Levelling, pp. 143, 144.

EIDOGRAPH

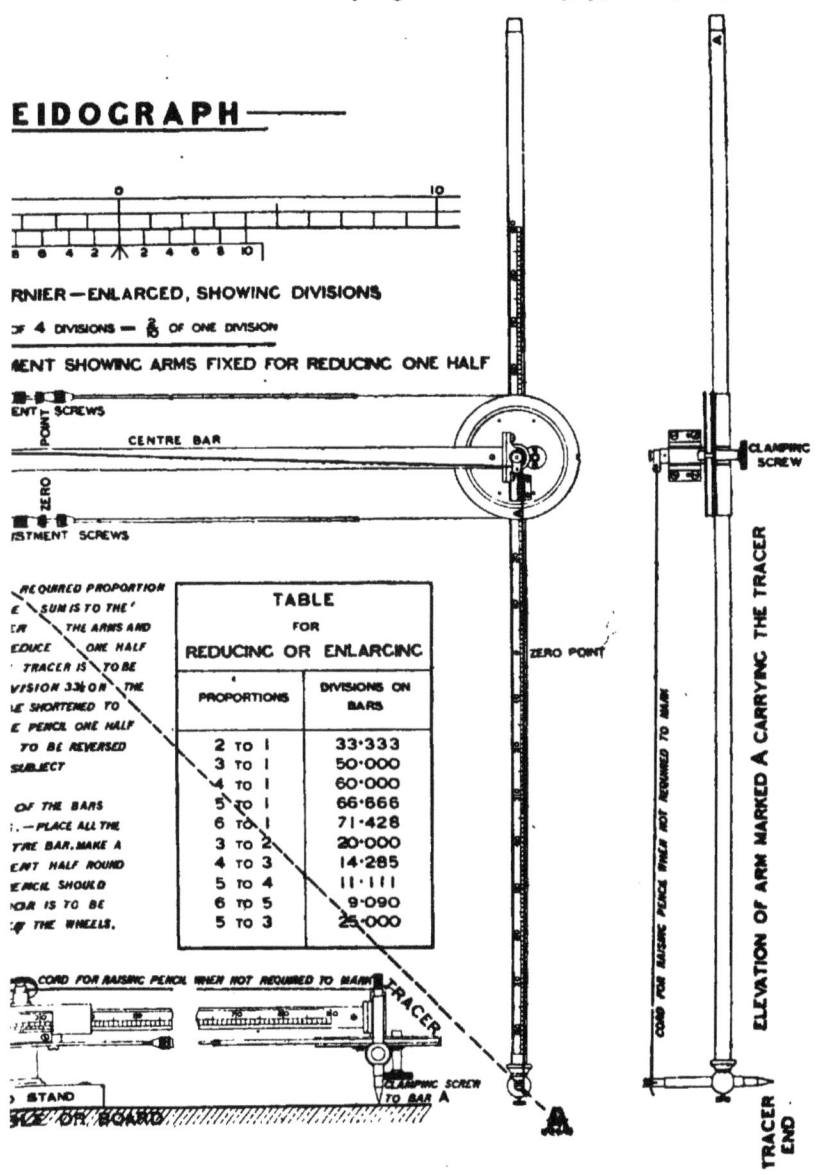

whole instrument is solely supported by the central fixed stand over which it works. Thus the friction upon the surface of the paper is reduced to a minimum, and there is less vibration than with the use of the comparatively thin arms of a pentagraph. When employed for reduction, as shown in the detail plan (pages 143, 144), the portion of the central bar carrying the bar marked A is heavier than the portion of the central bar carrying the bar marked B. The bar marked A carries the tracer and the bar marked B carries the pencil. The central bar is clamped at C. Hence the balance-weight is placed upon the short end of the central bar to steady the instrument in such a position that the whole is balanced over the fulcrum under C.

The geometrical construction of these intruments is based upon the principle involved in the construction of similar triangles. In both the pentagraph and the eidograph the pencil-holder, the tracer, and the fulcrum should under all circumstances be in a right line when set up ready for use, so that a fine string stretched from the pencil-holder to the tracer-holder should pass over the fulcrum as indicated by the dotted line shown in the diagram (pages 143, 144) which illustrates the eidograph. Otherwise the instrument is not correctly set, and must be re-adjusted.

Comparing the two instruments, the eidograph can cover a greater surface of reduction than the pentagraph. There is no sensible friction upon the single fulcrum centre of support, but to steady the instrument some makers add a movable castor, the roller of which is 2 inches diameter, which can be attached to the main beam with its axis in a direct line to the fulcrum of the central stand when the instrument is to be employed for reduction below one-third. The motions of the pencil and tracing point in a pentagraph are each derived from two circular motions, one about the joints at the ends of the arms, to which they are attached, and the other about the fulcrum. The radii of these travelling movements form the sides of similar triangles, of which the dotted line, passing through the pencil, the tracing point and the fulcrum form the third side. The distances traversed by the pencil and tracing point will maintain a constant ratio to one another when the

ENLARGING AND REDUCING PLANS. 146

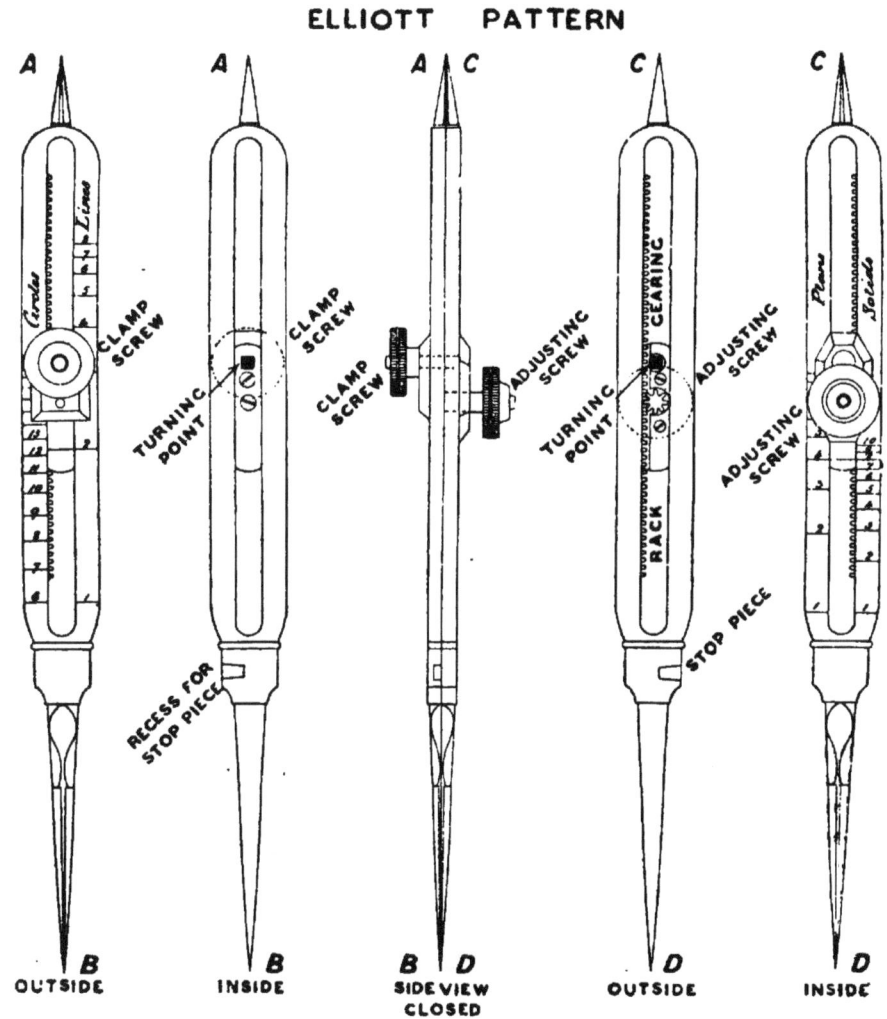

instrument is properly set, due to either of the above circular motions, and will therefore have also the same ratio when the two motions are combined. The inscribed space bounded by the arms of the instrument, as stated above, form a true parallelogram, and the proportion of the sides depends upon the setting of the instrument. In an eidograph the pulley-wheels under the ends of the centre bar are of exactly the same diameter, and the wheels are caused to move simultaneously by means of the steel bands attached to them. The tension of the bands can be adjusted by the screw connections when needed. Rules for setting the instrument will be found upon the diagram (pages 143, 144).

The pencil-holder is made to slide easily up and down the cylinder in which it rests, and the draughtsman is enabled to raise it off the paper when not required to mark by gently pulling the silk cord attached to the cranked lever arm, which cord passes over the bars of the instrument to the tracer point to which it can be fastened if desired, so that the draughtsman can prevent false or unnecessary marks being made upon the paper by passing over the cord at the tracer end two fingers of the same hand that moves the tracer. Additional small weights are provided to rest upon the top of the pencil-holder when the pencil is required to make strong marks upon the paper.

When the plan to be reduced contains numerous buildings the use of the proportional compass will be found advisable.

The plan to be reduced, and the paper upon which the reduced plan is to be plotted, are covered with squares drawn to their correct scales respectively, and the intersection of the lines is indexed for reference by numbers in the one direction, and by letters in the other direction. Proportional compasses fitted with a rack-gearing for the movement of the slider are the best, as they can be most accurately set by means of the milled-headed adjusting screw shown in the diagram (page 146), and when clamped by the opposite milled-headed screw are not liable to slip. To move the adjusting screw, the clamp screw must be first loosened, and the instrument is then so set that when the arms are opened, the distance between the points marked A and C bears the required proportion to the distance

ENLARGING AND REDUCING PLANS.

between the points marked B and D. As shown in the diagram (page 146), the body of the instrument consists of two narrow flat pieces of metal, each having a groove up to the centre, and united by a pair of slide pieces fitted into the grooves, and connected by a pin which traverses the axis of the instrument, and can be clamped in any required position by the milled-headed screw. A steel point is attached at both ends of each arm.

The scale of *lines* shows the marks at which the index must be set for so fixing the compasses when closed that the distance between the steel points when opened at one end shall bear a definite proportion to the distance between the steel points at the other end. This scale is only available where the ratio can be expressed by some whole number divided or multiplied into one unit, and cannot be applied so readily to ratios of 2 to 3 or 3 to 4, which are very often required.

The scale of *circles* will divide the periphery of a circle into any number of equal parts up to 20. The slide is set to the number of divisions required, then the points of the long arms of the instrument are opened to the radius of the circle and the distance between the points of the short arms will then indicate the chord to be taken upon the arc of the circle which divides the circumference according to the setting. The scale of plans applies to areas. A circle struck with the long points when the index on the slide is set to any registered number, will describe a circle the area of which will be exactly that registered number of times the area of a circle set out by a radius equal to the distance between the points of the short arms of the instrument. The scale of solids is similarly applied to cubical dimensions, and gives the comparative proportions for relative capacities. (See page 146).

CHAPTER XIV.

COMPUTING SCALES AND PLANIMETERS.

Computing scales made in boxwood are generally kept in stock for application to plans drawn to the Ordnance scale of either $\frac{1}{2500}$, $\frac{1}{500}$, or of 6 inches to 1 mile, but they can be made to any scale. Makers, however, generally require a week's notice for a computing scale different from their stock sizes, as they find the planimeter is so rapidly superseding the use of a computing scale that it does not pay to keep various sizes of this scale in stock. The annexed illustration shows a computing scale made to the Tithe Commission pattern of 3 chains to 1 inch (page 150). The large figures in the portions of the accompanying samples of computing scales denote acres, and the sub-divisions numbered 1, 2, 3 indicate roods. A projecting frame having a fine wire drawn across its centre is attached to a scale admitting of sliding motion in the direction of its length. The perches are engraved upon the ivory scale attached to this movable metal frame, the use of which will be understood upon reference to the annexed drawing (pages 151, 152), showing a complete instrument. The application of a plan is explained in note 1. Four scales are shown upon this instrument. (See note 2.) The example given supposes the plan to be drawn to a scale of 4 chains to 1 inch, in which case the calculated distance between the parallel lines upon the tracing paper is seen to be $\frac{1}{4}$ inch, which is ruled off in the direction of the greatest length, in order to have as few strips as possible and form a gridiron tracing of parallel lines $\frac{1}{4}$ inch apart. The number of square chains in 1 inch upon the plan, divided by the number of lineal chains to an inch upon the scale, will give the number of parallel lines or spaces to be ruled within a depth of 1 inch upon the computing paper. Thus, with the use of such tracing strips, when the upper scale marked 1 is to be employed, a length of $2\frac{1}{2}$ inches is seen to measure an acre. The object of the calculation given in note 1 upon the diagram for the determination of the value of X, is to arrive at the actual width required to suit the scale of the plan for recording an area of 1 acre for an actual length of rectangle registered as an acre upon the computing scale. Thus, referring to note 2, the length upon the scale of $12\frac{1}{2}$ inches records 5 acres in the case of scale 1, because one-fifth of $12\frac{1}{2}$ inches multiplied by $\frac{1}{4}$ inch = $\frac{5}{8}$ square inches, and since by the scale of the plan 16 square chains equal 1 square inch, an area of paper $\frac{5}{8}$ square inches equals 10 square chains or 1 acre. In the case of scale 2, we have the same total length of $12\frac{1}{2}$ inches, reading two acres, and

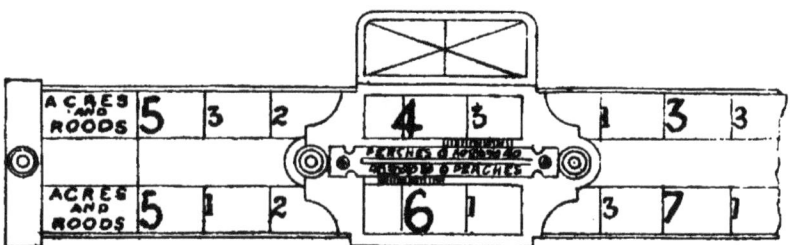

hence the divisions upon the lines for the tracing paper would need to be $\frac{1}{10}$ inch apart, which would not be so suitable as scale 1 for this special plan. The wire line C in the frame A B is first so set that the frame rests against the stop-piece F. It is then placed upon the tracing paper over the area to be calculated so as to start from zero at the line M, with the edge of the scale parallel to the lines upon the tracing paper. The scale is maintained in position by the pressure of the hand, and after carefully moving the frame so that the line C traverses from M to N the instrument is then lifted up and replaced with its edge parallel to the lower lines upon the tracing paper, so that the wire C starts from O at the same position on the scale as it indicated when at N. The frame is traversed over each rectangle successively from M to N, O to P, Q to R, S to T, &c., by

COMPUT

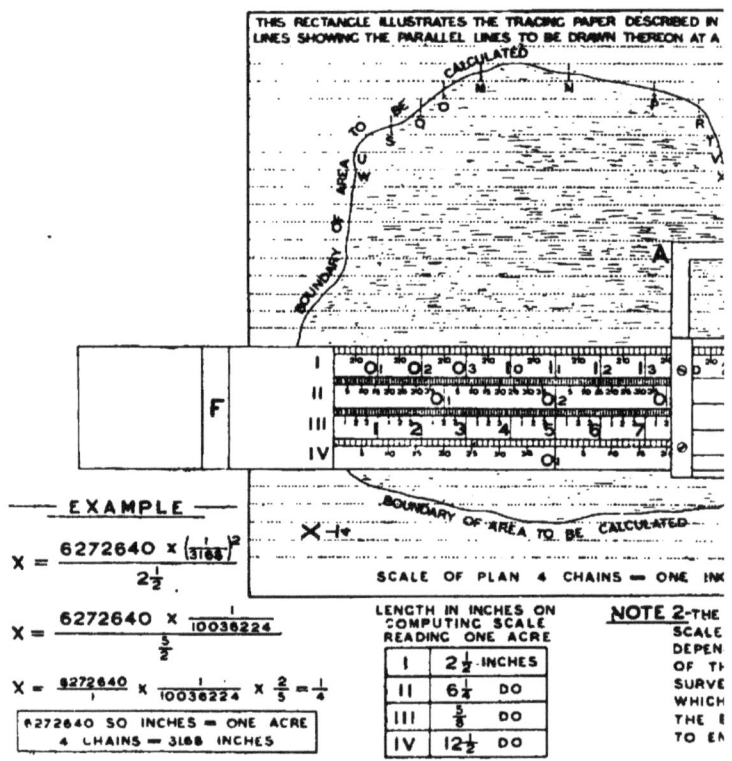

Land Surveying and Levelling, pp. 151, 152.

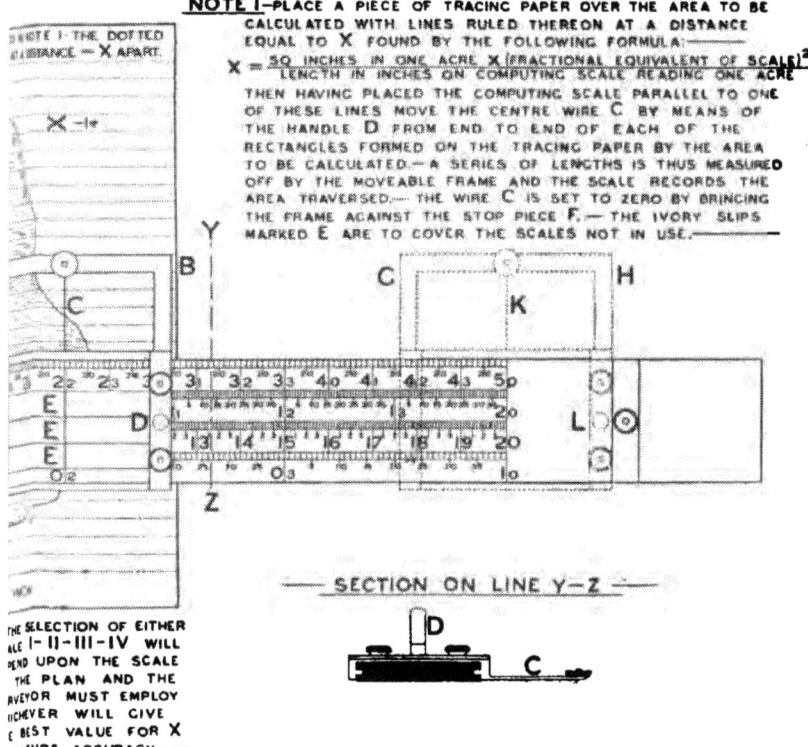

means of the handle D (see section upon the line Y Z), the exact position of the starting and finishing points for each rectangle being judged by the eye so as to be equal to and to compensate one another in the give and take, and thus by a series of mechanical additions the area can be read off the computing scale in acres, roods and poles. With the use of the upper scale, as shown, when the frame reaches the stop-piece L it indicates that five acres have been traversed.

In Merrett's improved computing scale the screw in the metal frame is made to act as a clamp, and different scales are supplied to fit the same metal frame, instead of the ivory pieces employed to cover the scales which are not required for use in the instrument adopted by the Tithe Commission Office. The reading edge is also bevelled off against the scale. It is evident that the degree of accuracy

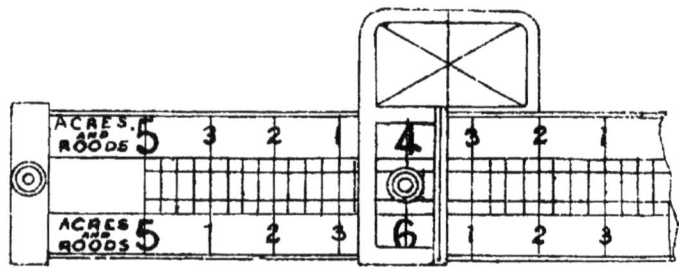

obtainable will be proportionate to the skill of the surveyor in the equalisation of the extremity of each irregular boundary to a space traversed by the scale.

It is customary in this country to express the area of land in acres, roods, and poles, except such land as is portioned off for building purposes, which it is usual to compute in superficial yards. The former area may easily be ascertained by the use of a computation scale, but precaution must be observed to keep the tracing paper fixed to prevent it shifting, and, with due care, accurate results can then be obtained without any further calculation being necessary, in which the number of chains traversed is recorded upon this instrument by a sliding vane in a purely mechanical manner.

The annexed figure shows how a triangle A F H may be drawn equal in area to a five-sided figure A B C D E, the area of which triangle is easily arrived at by the

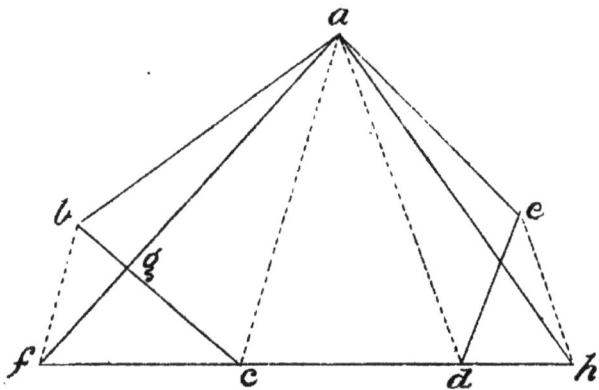

simple rules of mensuration. But the methods frequently employed by unskilful surveyors to find the area by offsets are erroneous, when they divide the sum of the offsets by their number for a mean breadth, or, as some do, in dividing that sum by one more than their numbers for a mean breadth. The first of these methods gives an area too little, and the second an area sometimes too little and sometimes too much, when such mean breadth, as thus found, is multiplied by the whole base for arriving at the superficial amount. A third method, which is usually more accurate than either of the preceding ones, is to set down each offset twice (accounting that one where the boundary meets the station line), except the first and last, which are only entered once. The sum of these offsets is then multiplied by the base, the product divided by the number of offsets set down, and the quotient given as the area required.

The reference numbers upon Ordnance maps (see pages 130 and 240) enable a book of reference to be compared with any enclosure. The areas upon these maps have been computed by the Ordnance Survey Department to the centre of the fence or other boundary of the enclosure, except in the following cases:—

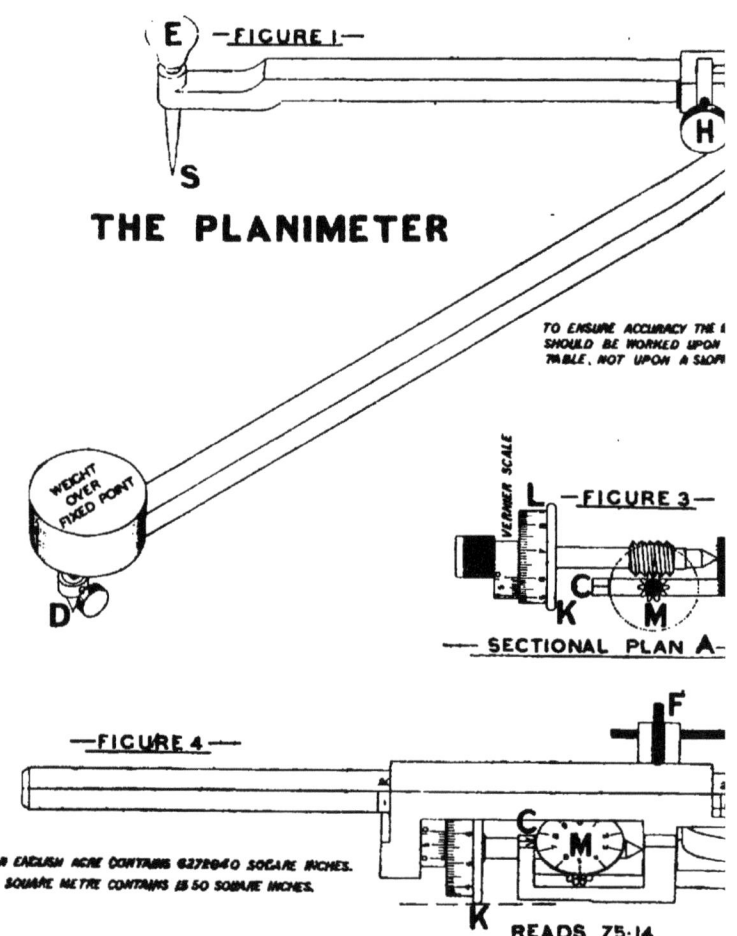

Land Surveying and Levelling, pp. 155, 156.

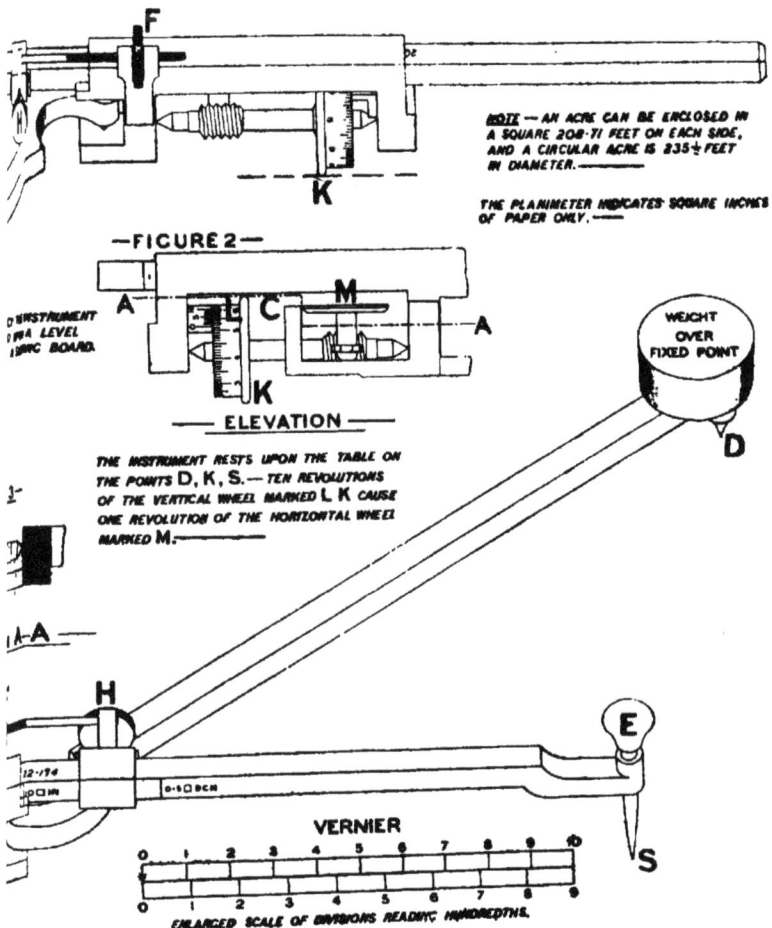

(1) When the fence or other boundary is also the boundary of a parish or other civil division which does not follow the centre of the fence, the area is calculated to the parish or other boundary, and not to the centre; (2) when the fences, &c., bounding either side of a railway are included wholly within its area.

In the planimeter (see pages 155, 156) the bar E F can be made either to slide in the box to which the other arm is attached (see fig. 1) or can be permanently fixed to this box. In the latter case it is constructed to record square inches only, and is known as the fixed scale planimeter. In the former case it will record square inches when the mark upon the sliding box is made to come immediately under the mark upon the bar E F, which is situated near the figures 22·174 in fig. 4, and is distinguished as the proportional planimeter. The sliding box is fixed to the sliding arm E F by the clamp screw H, and the two marks above alluded to are made to coincide by the movement of the slow-motion screw F. The horizontal wheel M records the square units in one revolution, the wheel L records square units and tenth parts of a square unit, and the vernier attached outside the wheel L enables hundredths of a square unit to be read. When the needle-point D is placed outside the area to be calculated no account is taken of the figures upon the sliding bar E F, but when the needle-point D is placed inside the area to be calculated, the number engraved upon the bar is to be added to the reading of the instrument after the boundary of the area has been completely traversed by the pointer S, before the first reading of the instrument is deducted.

Goodman's Patent Planimeter is an economical form of instrument, and may be thus explained :— Let it be required to measure the area of the figure, page 158. Choose a point A, as near the centre of the figure as can be judged by eye, and from it draw a line A B to the boundary. Hold the tracing leg of the instrument in the right hand, placing the point at A and the hatchet at X (*i.e.*, with the instrument roughly square with A B, see fig. 3), and press the hatchet in order to make a slight dent in the paper at X; then, the finger having been removed from the hatchet, the tracing point of the instrument is caused to traverse the line A B and the boundary in the direction indicated by the arrows, returning to A via A B, when it will be found that the hatchet has taken up a new position and it must be again lightly pressed (as illustrated in fig. 2) in order to make a fresh dent in the paper at Y (fig. 3). The instrument being held in this position, revolve the paper on which the figure is drawn through about 180° (by eye), using the point of the instrument as a centre (as shown in fig. 4), and taking care that neither the point nor the hatchet shift while the paper is being turned. The line A B will again be roughly at right angles to the axis of the

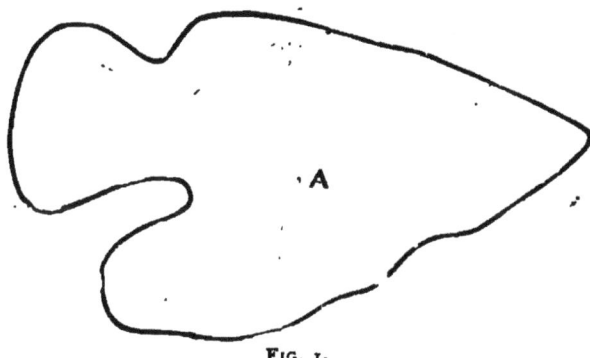

FIG. 1.

instrument, but in a reversed position (see dotted figure, fig. 3). Now cause the tracing point to traverse the boundary as before, but in the opposite direction, as indicated by the dotted arrows. The hatchet will take up the new position X_1 which may or may not coincide with X; then, the mean of X Y and X_1 Y measured on the scale engraved on the instrument is the area of the figure; this can be readily read off by pricking a central point as shown between X and X_1 by eye. When it is inconvenient to turn the paper round, the instrument itself may be turned round to form a dent X_1 on the opposite side of the figure, as shown at fig. 3a. Then by following the boundary in the direction of the arrows Y^1 is obtained. The area is the mean of the lengths X Y and $X_1 Y_1$ measured off on the scale as before, or the area $= \dfrac{X Y + X_1 Y_1}{2}$

When the area is large, the instrument will move through a large angle, and consequently, if approximately square with A B at starting, it will be a long way out at the finish. In such a case all that is necessary is to see that the *mean* position of the instrument is square with A B.

The following precautions are to be observed:—

Do not allow the hatchet to work on a rough surface—

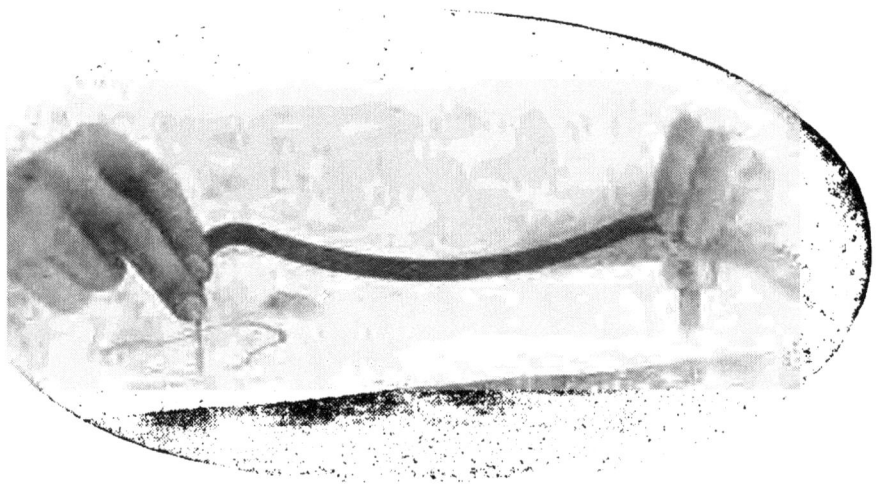

Fig. 2.

smooth writing or drawing papers are suitable; do not work on wood, as the hatchet tends to travel along the grain.

Do not allow the hatchet to go off the edge of the paper or over ridges. Hold the instrument freely so that the motion of the hatchet shall not be interfered with. It is always advisable to use the weight on the hatchet to prevent side slip.

Use the instrument on a flat table, not on a sloping desk.

On no account attempt to sharpen the hatchet, either with an oilstone or otherwise.

See that the instrument is held with its legs fairly vertical.

In measuring the area off always work from the zero of the curved scale.

This instrument is an improvement on the Stang planimeter made by Knudsen, of Copenhagen. In a pamphlet published by him he gives the complete mathe-

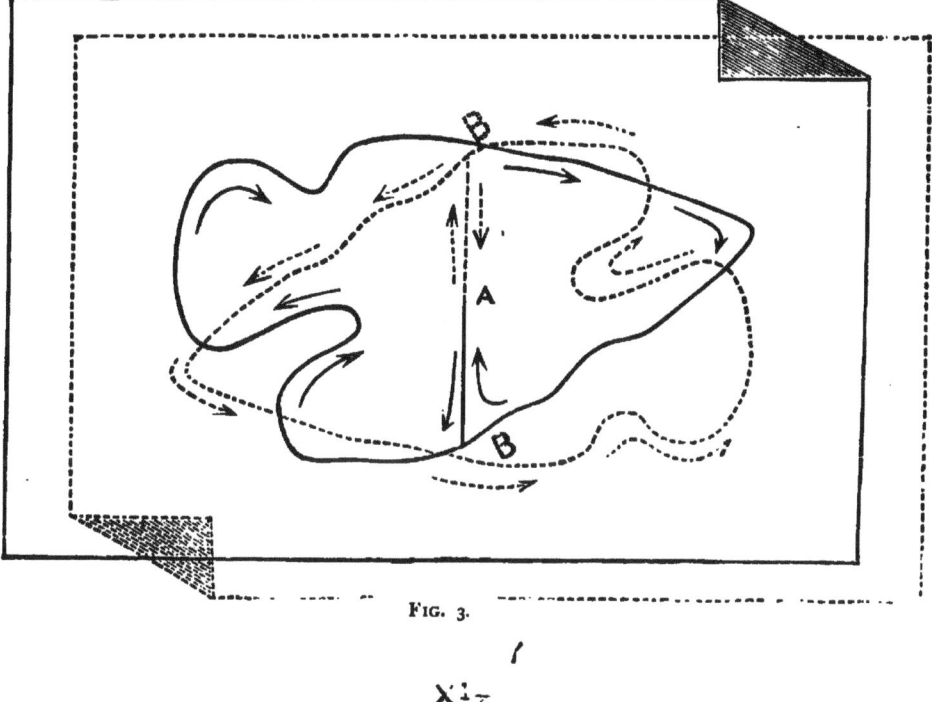

Fig. 3.

matical theory of the instrument and arrives at the following result :—

$$I = \frac{C_1 + C_2}{2} p \left[1 - \left(\frac{R}{2p}\right)^2\right]$$

Where I = the area traced out by the pointer in sq. inches,
C_1 = the distance between the dents X and Y in inches,
C_2 = ,, ,, ,, X_1 and Y ,,
p = the length of the instrument in inches,
R^2 = the mean square of the radii of the figure
(Le milieu des carrés des distances de T (a) jusqu'à la circonférence)

LAND SURVEYING AND LEVELLING.

The making of such a calculation for every area measured is far too tedious an operation. By using Goodman's instrument as described above, first in one direction and then in the other, the quantity $\frac{C_1 + C_2}{2}$ is obtained by a direct reading. If the instrument were of very great length as compared with the dimensions of the area, the quantity in the brackets would vanish, then putting $\frac{C_1 + C_2}{2} = c$, the area would simply be equal to the product $p\,c$, and a scale for measuring the distance c in which 1 square inch was

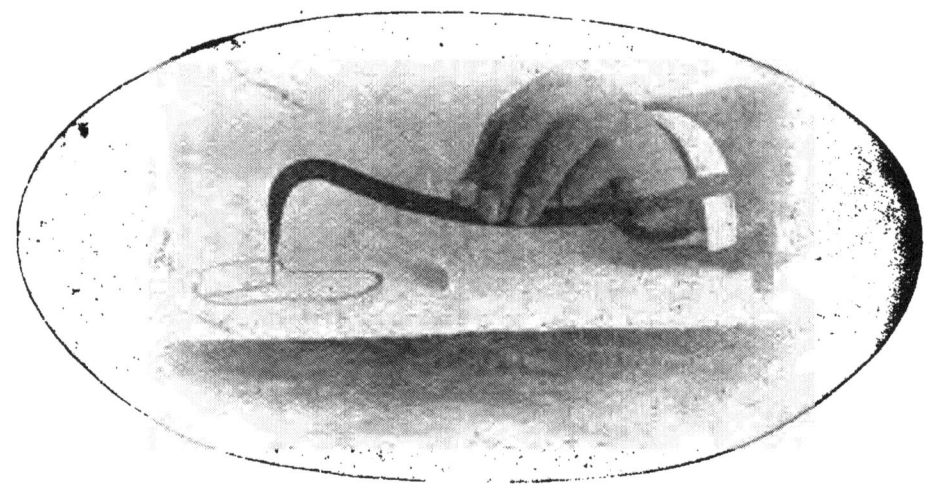

FIG. 4.

made equal to $\frac{1}{P}$ inches would at once give the area in square inches. But as the areas generally dealt with are of somewhat large dimensions as compared with the length of the instrument, the quantity in brackets must not be neglected. Assume for a moment that we are measuring the area of a circle with the instrument—its area is proportional to R^2, hence as $(2p)^2$ is a constant for any given instrument, the whole quantity $(\frac{R}{2p})^2$ is proportional to the area traced out by the pointer. Thus by making a scale with gradually increasing divisions this quantity may be entirely eliminated. This is what is accomplished in Professor Goodman's planimeter, and instead of having to solve the equation as given above in order to find the area, it is read off direct from the scale without any calculation.

If the area dealt with is not a circle, the error involved in assuming that its R^2 is equal to the R^2 of a circle of equal area is so small that it is quite inappreciable on a scale which only reads to tenths of a square inch, it would indeed seldom be appreciable on a scale reading to hundredths of a square inch.

Planimeters are now so well understood, and their results are found to be so reliable, that no Surveyor's office is complete without them. A level table is essential to ensure accuracy. The instruments shown upon pages 155, 156 are known as Amsler's planimeters. An example of their employment is furnished in the chapter on Contours (pages 197-199).

CHAPTER XV.

TAKING LEVELS.

THE chief use of instruments for taking levels is to compare the heights of different stations with reference to a fixed datum or horizontal line. The datum may be the sea level or any arbitrary horizontal line that can be localised. The method of procedure is virtually the same in all cases. An accurate plan is absolutely essential, and the surveyor has first to obtain a detailed map of the country upon which the line he is to follow is marked, or else he must make a plan for himself in order to keep a record of any line that may be marked upon the ground, or otherwise indicated. Hence surveying operations precede those of levelling. Hutton, an eminent mathematician who was created a Fellow of The Royal Society in 1774, and wrote at the end of the eighteenth century upon mathematical subjects, states "that two or more places are on the same level when they are equally distant from the centre of the earth. Also one place is higher than another, or above the level of it, when it is further from the centre of the earth, and a line, equally distant from that centre in all its parts, is called a line of true level."

Levelling.—The field work connected with the process of taking levels may be illustrated by the diagram headed "flying levels," that is, levels in which a description is not required of *all* the intermediate points upon which the staff is to be held in order to connect the work. (See pages 166-168). Suppose the level of the point A is known by previous levelling to be 60 ft. above a given datum, and that it is simply required to ascertain the level of the point H. In the first place, attention should be given to the adjustments of the instrument employed. The verticality of its axis, and consequently the horizontal position of the azimuth, must be proved. It is furthermore necessary to remove the instrumental parallax, and to see that the line of collimation is correct, so that when the telescope is set level, parallel to the optical axis, and to the surface of the cylindrical rings upon which it is supported, the cross wires in the diaphragm shall coincide with the axis of the supports in which the telescope rests. The distance apart, as well as the uneven nature of the surface of the ground between the points A and H, render it necessary to call upon intermediate points, such as B, C, D, E, F, and G, and to re-set up the instrument between each change of position in the level staff. The instrument is set up level between A and B; the staff is placed upon A, and found to read 2·20. This is booked as a *back-sight*, because we are looking back upon a point the level of which is already known. The telescope is then revolved horizontally; the staffman is directed to move forward to B, or to a point where the observer can read as great a height upon the staff as possible, owing to the falling inclination of the ground. The staff at B is booked as a *fore-sight*, because the observer is here looking forward to a point the level of which he does not yet know. Having read B, he then moves the instrument forward, while the staffman is entrusted to maintain the foot of the level-staff very carefully upon the point B. To assist the staffman in so doing, he generally employs a foot-plate or iron peg and waves the staff upon it to and from the telescope, like an inverted pendulum over a vertical line, or line normal to a tangent to the earth's surface at the spot where the level staff is being held when the observer is reading the divisions. With the instrument in the new position between B and C, placed so as to read as low down as practicable upon the staff, the sight 1·11 is then read. This is booked as a back-sight, because when the level book is reduced, the level of the point B can be arrived at before the sight 1·11 is considered. Suppose the ground to be such that the instrument is placed upon the left hand of A, when the sights 2·20 and 12·09 are read, or upon the left hand of B, when 1·11 and 11·70 are read with the telescope pointing in the same direction, the first sights taken with the instruments would still be booked as " back-sights " for the reason above stated, although the observer would be looking forward, in each case, so far as direction is concerned.

The position of the level staff is localised, not that of the level. The level is simply set up in any convenient position for reading the staff, and the surveyor should never hesitate to take an extra sight, off his line of section, if need be, as a fore-sight, in order to secure a firm position for the level staff while changing the position of the level. The intermediate spots upon which the staff is held between A and H can be described, if required, by letters referring to an accurate plan, from which the horizontal distances can be afterwards measured with a scale for the purposes of plotting, or the whole can be chained in the field as a continuous section.

Where the ground is undulatory, and an accurate vertical section is required, it will be found necessary to take some sights between the first and last, with the level in any single position. These are called intermediate sights, and the entries are recorded in a separate column. (See pages 173-176.)

In the diagram of "Flying Levels," if the back-sight at A reads 2·20, the level of the line of collimation would be 62·20, and the level of the ground at B would be found by subtracting the fore-sight 12·09 from 62·20, giving a result of 50·11. Again, adding the back-sight 1·11 to 50·11, we obtain the level of the new line of collimation between B and C, equal to 51·22, and subtracting the fore-sight 11·70 we obtain 39·52 as the level of the ground at C. Thus, by continuously adding the back-sights and subtracting the fore-sights taken between A and H, we may obtain the reduced level of the point H as 39·42. This will be clear upon reference to the diagram (pages 166-168).

Now, if instead of first adding 2·20 to 60·00, and subtracting 12·09, we took the difference between 2·20 and 12·09 and subtracted this difference from 60·00, we should obtain the same result, 50·11, independently of any calculation for the level of the line of collimation. It will be seen that this is exactly the usual method adopted in a level book, in which columns are provided for the amounts of rise and fall; the difference constituting *a rise* when the

Land Surveying and Levelling, pp. 166, 167, 168.

NOTE—THE FIRST SIGHT READ OFF THE LEVEL STAFF IS BOOKED AS A BACK SIGHT, AND GIVES THE HEIGHT OF THE LINE OF COLLIMATION IN THE INSTRUMENT, ABOVE THE POINT ON WHICH THE LEVEL STAFF IS PLACED.———

THE LAST SIGHT READ OFF THE LEVEL STAFF BEFORE CHANGING THE POSITION OF THE INSTRUMENT, GIVES THE VERTICAL DISTANCE BETWEEN THE LINE OF COLLIMATION, FOUND AS ABOVE, AND THE POINT ON WHICH THE LEVEL STAFF IS THEN HELD.——THIS SIGHT IS BOOKED AS A FORE SIGHT, AND THE LEVEL STAFF IS REPLACED ON THE SAME POINT AS THAT UPON WHICH THIS LAST FORE SIGHT IS TAKEN, FOR ASCERTAINING THE FIGURES TO BE BOOKED, WHEN READING THE NEXT BACK SIGHT, IN CONTINUING THE LEVELLING ——

LUMNS, GIVE
D.

MNS, GIVE
E.

C, D, E, F, G, H
LD BE

60.00
20.58
―――
39.42

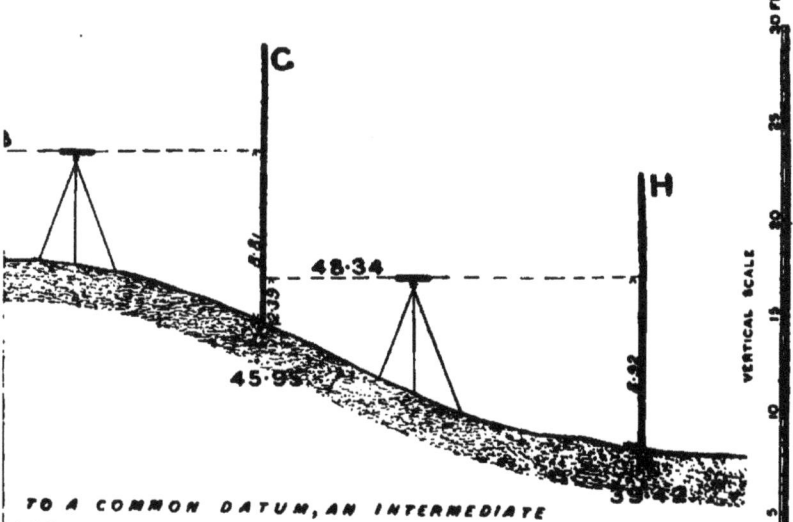

TO A COMMON DATUM, AN INTERMEDIATE
GHT TO THE SIGHT BOOKED IN THE LINE
SIGHT TO THE SIGHT BOOKED IN THE LINE
IFFERENCE BETWEEN THE SUM OF THE
M OF THE FORE SIGHTS, IS EQUAL TO THE
TOTAL RISE AND FALL; AND ALSO TO THE
LAST AND FIRST REDUCED LEVEL.——

fore-sight is less than the back-sight, as between D and C, or E and F or *a fall* when the fore-sight is greater than the back-sight, as between A and B, or C and D. (See page 173.)

Level Book—The annexed pages (173-177) give the readings taken upon a staff, in the first place, when adjusting for collimation, and assume that the circumstances are such that the collimation is 0.06 foot out of truth, showing that correction must be made, until the staff at W reads, in this case, 8·86. They also show the readings booked in taking certain check levels. The system of flying levels is adopted for running a check level after taking a section.

In the record of the longitudinal section taken on July 16, 1900, the student is recommended to copy into an actual Level Book the entries here provided, and then to reduce the levels page by page, there being four pages in this section to reduce.

All readings between the first and last sights before changing the position of the instrument are entered as intermediate sights. The first sight read off the level staff is booked as a back-sight, and gives the height of the line of collimation in the instrument, above the point upon which the level staff is placed. The last sight or reading read off the level staff before changing the position of the instrument is booked as a fore-sight. The columns of distances or lengths give the position of the level staff.

Horizontal distances are expressed in links, and are localised by description in the column headed Remarks, which usually occupies the whole of the right-hand side of each double page of the book.

The first back-sight is preferably taken on a bench mark.

The reading of each intermediate and fore-sight gives the vertical distance between the line of collimation and the points upon which the level staff is held. The surveyor observes the staff while it is being waved and books its lowest reading; he then should (if at all uncertain) look again at the staff to corroborate the observation, and if the record requires alteration, use a piece of india-rubber, which he should always have in his pocket ready for use, so as to keep his level book clearly and

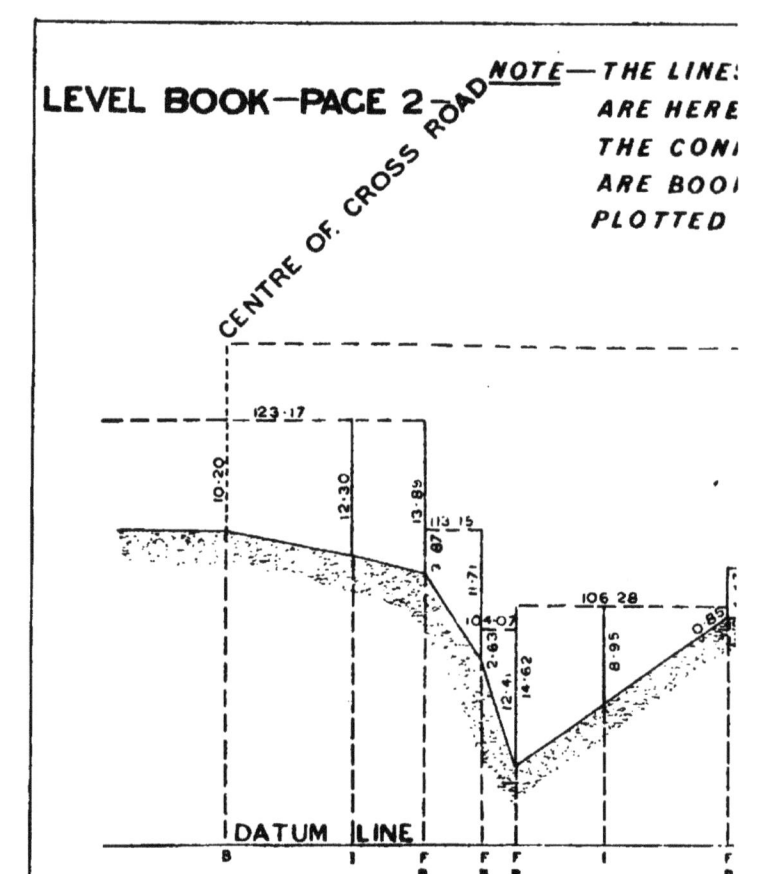

Land Surveying and Levelling, pp. 170, 171.

OF COLLIMATION, AT THEIR RESPECTIVE LEVELS
DRAWN, SOLELY FOR THE PURPOSE OF SHOWING
EXION BETWEEN THE STAFF READINGS, WHICH
ED IN THE FIELD; AND THE SECTION, WHICH IS
FROM THE REDUCED LEVELS. ———

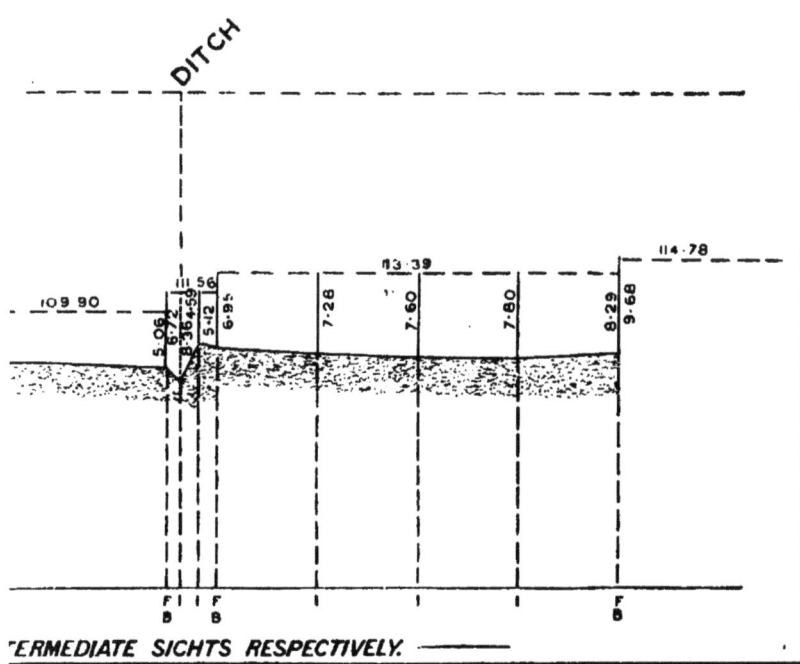

ERMEDIATE SIGHTS RESPECTIVELY. ———

distinctly. The columns headed Back-sight, Intermediate, Fore-sight, and Lengths, denote the entries made in the field, while the columns headed Rise, Fall, and Reduced Levels, denote the entries afterwards made in the office. Some level books place the column headed Rise in the first column, to distinguish it more particularly from the column headed Fall. Other books place the column headed "Distance" first. The columns marked Reduced Levels and Distances or Lengths are the columns used in plotting the levels. Sometimes the location of the level staff is referred to by a letter, without any measured distance or length being taken. A very accurate plan is in this case necessary to localise the description. When for purposes of merely rendering the levelling continuous no distances are needed to be recorded, and the location of the level staff is not required to be noted, it is better, rather than leaving a blank in the distance column, which might imply an omission, to draw a short dash across the space, which shows the point to be one that has been considered.

In reducing the levels to a common datum, an intermediate sight serves as a fore-sight to the sight booked in the line before it, and as a back-sight to the sight booked in the line following it. The difference between the sum of the back-sights and the sum of the fore-sights is equal to the difference between the total rise and fall, and also to the difference between the last and first reduced level. For temporary bench marks, take the highest point on the top of a fixed stone; nails driven into the trunk of a tree to serve as bench marks are not satisfactory. The letters B. M. at the top of the column for lengths, indicate a bench mark, the level of which is given in the column of reduced levels, and from which all the succeeding levels are calculated. In every case a full description of the position of a bench mark, accompanied, if necessary, by a sketch to show surrounding objects and the part of the structure on which it is fixed, should be made in the level book. All observations connected with either the distances, levels or general description of the ground passed over, should be entered in the level book at the time of noting the same in the field; nothing should be trusted to memory.

TAKING LEVELS. 173

Adjustment for Collimation, July 16th, 1900.

Back sight.	Intermediate	Fore sight.	Rise.	Fall.	Reduc'd Levels.	Lengths	REMARKS. NOTE.—This page is for descriptive remarks as to location.
10·98	...	...	...	...	...	W	
6·57	...	8·69	2·29	...	...	Z	
...	8·92	...	...	2·35	...	W	
...	...	8·86	0·06	...	...	W	
17·55	...	17·55	2·35	2·35	...	...	

Flying Levels between points A and H.

2·20	...	...	...	...	60·00	A	see pp. 166-168.
1·11	...	12·09	...	...	...	B	
2·30	...	11·70	...	...	...	C	
8·28	...	7·51	...	...	...	D	
11·00	...	1·90	...	...	...	E	
6·37	...	3·30	...	...	...	F	
2·39	...	8·81	...	...	...	G	
...	...	8·92	...	...	...	H	

Longitudinal Section, taken on July 16th, 1900.

Back sight.	Intermediate	Fore sight.	Rise.	Fall.	Reduc'd Levels.	Lengths	REMARKS. NOTE.—This page is for descriptive remarks as to location.
3·71	...	...	...	...	117 94	B.M.	on side of road in wall opposite lamp post near Half Way House.
...	5·06	...	...	...	...	0	at 139 feet from L P. shown on plan.
9·40	..	7·88	...	...	...	200	
...	10·20	...	...	...	...	375	centre of cross road.
...	12·30	...	...	...	...	400	
3·87	...	13·89	...	...	...	414	see pp. 170, 171, and pp. 178, 179.
2·63	...	11·71	...	...	...	426	
14·62	...	12·41	...	...	...	433	
...	8·95	...	...	...	...	450	
4·47	...	0·85	...	...	...	475	
...	2·99	...	...	...	...	T.B.M.	lower hinge of gate to Wood Lodge.
		5·06				510	

LAND SURVEYING AND LEVELLING.

LONGITUDINAL SECTION, taken on July 16th, 1900—*continued*.

Back sight.	Intermediate	Fore sight.	Rise.	Fall.	Reduc'd Levels.	Lengths	REMARKS. NOTE.—This page is for descriptive remarks as to location.
6·72	...	...	...	...	...	...	at 510.
...	7·25	...	...	...	...	...	on edge of ditch.
...	8·36	...	...	...	...	512	bottom of ditch.
...	4·59	...	...	...	...	516	stump on top of bank.
6·95	...	5·12	...	...	...	520	on fair ground.
...	7·28	...	...	...	...	540	
...	7·60	...	...	...	...	560	
...	7·80	...	...	...	...	580	
...	...	8·29	...	...	...	600	
9·68	...	...	...	...	...	600	
7·54	...	1·66	...	...	...	625	fence crosses at 624.
...	5·66	...	...	...	...	650	centre of turnpike road.
...	6·53	...	...	...	...	680	fence crosses at 676.
10·12	...	2·16	...	...	...	721	
6·33	...	9·62	...	...	...	750	
4·41	...	13·53	...	...	...	775	
7·88	...	13·51	...	...	...	800	
0·12	...	12·08	...	...	...	825	
...	9·77	...	...	...	...	850	
1·35	...	14·12	...	...	...	860	
...	5·95	...	...	...	...	880	fence crosses at 882.
2·61	...	10·85	...	...	...	900	
...	8·11	...	...	...	...	918	edge of river. Wa'er level at 1.30 p.m.
...	13·51	...	...	...	...	933	bottom of river.
...	8·11	...	...	...	...	950	edge of river water level (opposite side).
13·75	...	1·46	...	...	...	967	
...	8·90	...	...	...	...	990	fence crosses at 993.
··	4·40	...	...	...	...	1,000	
...	4·18	...	...	...	...	1,020	

TAKING LEVELS.

LONGITUDINAL SECTION, taken on July 16th, 1900—*continued*.

Back sight.	Inter-mediate	Fore sight.	Rise.	Fall	Reduc'd Levels.	Lengths	REMARKS. NOTE.—This page is for descriptive remarks as to location.
13·27	...	0·08	...	...	...	1,050	
11·19	...	3·09	...	...	...	1,075	
10·54	...	3 69	...	...	...	1,100	peg at beginning of curve (radius 20 chains).
13·57	...	0·84	...	...	...	1,125	
...	6·97	...	...	...	...	1,150	
13·13	...	0·77	...	...	...	1,175	
13·82	...	3·03	...	...	...	1,200	
12·99	...	2·92	...	...	...	1,225	
...	7·59	...	...	...	...	...	intersection peg A.
11·18	...	12·78	...	...	...	1,250	
10·50	...	4·89	...	...	...	1,264	$10 \times \cot \frac{1}{2} (171°7'-12') = 155$ links
9·88	...	2·71	...	...	...	1,290	
7·16	...	2·77	...	...	...	1,315	fence crosses at 1,316.
5·39	...	11·46	...	...	...	1,340	
2 11	...	8·99	...	...	...	1,360	
1·71	...	10·61	...	...	...	1,385	
...	7·93	...	...	...	...	1,410	peg at end of curve.
1·30	...	13·52	—	...	...	1.454	
...	7·71	...	...	...	...	1,500	
...	...	12·15	...	...	...	1,525	
2·97	...	...	...	...	...	1,525	
1·77	...	11·40	...	...	...	1,550	
2·01	...	14·57	...	...	...	1,575	
3·59	...	4·61	...	...	...	1,592	edge of coping 9 in. × 6 in.
...	9·09	...	...	...	...	1,600	centre of stream at bottom
...	4·19	...	...	...	...	...	water level at 3.15 p.m.
...	3·56	...	...	...	...	1,608	edge of coping on opposite side.
...	4·09	...	...	...	...	1,620	fence crosses at 1,616.
...	4·99	...	...	...	...	1,638	centre of road (no footpath). Width on the square, 28 ft.
...	7·78	...	...	...	...	1,666	
8·12	...	12·89	...	...	...	1,675	fence crosses at 1,668.
...	12·62	...	...	...	...	1,690	
...	9·42	...	...	...	...	1,700	
...	14·09	...	...	...	...	1,725	
6·57	...	10·73	...	...	...	1,750	
...	11·53	...	...	...	...	1,880	
...	...	7·75	...	...	...	B. M.	(96·41) on gate-post by lodge to Rydal Villa.

176 LAND SURVEYING AND LEVELLING.

CHECK LEVELS, July 17th, 1900.

Back Sight.	Intermediate	Fore Sight.	Rise.	Fall.	Reduc'd Levels.	Lengths	REMARKS. NOTE.—This page is for descriptive remarks as to location.
2·99	...	...	...	...	117·94	B.M.	on wall, opposite L.P., near Half-way House (see page 2).
9·01	...	9·19	...	...	...	...	
3·79	...	13·83	...	...	...	lower	hinge of gate to Wood Lodge.
3·02	...	14·01	...	...	...	...	
1·81	...	14·82	...	...	...	...	
9·87	...	13·16	...	...	...	...	
...	13·82	...	...	...	...	water	level in river at 12 noon.
10·81	...	4·91	...	...	...	...	
12·62	...	0·00	...	...	...	...	
10·59	...	1·94	...	...	...	...	
12·78	...	3·09	...	...	...	peg	at beginning of curve.
13·23	...	1·87	...	...	...	...	
12·99	...	2·64	...	...	...	...	
11·60	...	1·89	...	...	...	...	
10·72	...	3·38	...	...	...	...	
13·19	...	2·63	...	...	...	inter	section peg A.
5·22	...	11·00	...	...	...	...	
13·01	...	14·05	...	...	...	peg	at end of curve.
2·11	...	14·57	...	...	...	...	
...	...	13 62	...	...	...	...	see pp. 182, 183.
4·19	...	...	...	...	...	...	
3·11	...	12·99	...	...	...	...	
3·62	...	14·31	...	...	...	...	
...	10·80	...	...	...	...	edge	of coping nearest road.
...	11·33	...	...	...	...	water	level 2.0 p.m.
1·25	...	12·60	...	...	...	...	
...	6·99	...	...	...	...	...	
...	...	12·57	...	...	...	B.M.	(96·41) on gate-post by lodge, Rydal Villa.

Note.—The transverse lines in the above examples show where the division of pages in an ordinary level book transpire.

TAKING LEVELS.

Contour Levels, taken July 19th, 1900.

Back Sight.	Intermediate	Fore Sight.	Rise.	Fall.	Reduc'd Levels.	Lengths	REMARKS. NOTE.—This page is for descriptive remarks as to location.
2·61	...	...	...	...	137·86	B.M.	on mile stone 4 m. to Tetbury.
1·83	...	12·85	...	10·24	127·62	...	
...	9·45	...	...	7·62	120·00	Contour	
6·78	...	9·31	0·14	...	120·14	...	(see Chapter XVII.).
...	6·92	...	...	0·14	120·00	...	
8·29	...	6·98	...	0·06	119·94	...	
...	8·23	...	0·06	...	120·00	Contour	
5·61	...	7·99	0·24	...	120·24	...	
...	5·85	...	...	0·24	120·00	...	
...	...	5·92	...	0·07	119·93	top of	peg fixed in the field side of the hedge adjoining private road to the farm house.
25·12	...	43·05	0·44	18·37 ⎫	17·93		
...	...	25·12	...	0·44 ⎭			
		17·93		17·93	137·86		

Note.—For plan and description see pages 194, 195.

For description of Ordnance Datum see page 139, of Trinity High Water see page 140, and of a datum upon a parliamentary section pages 245-247 and 269.

H

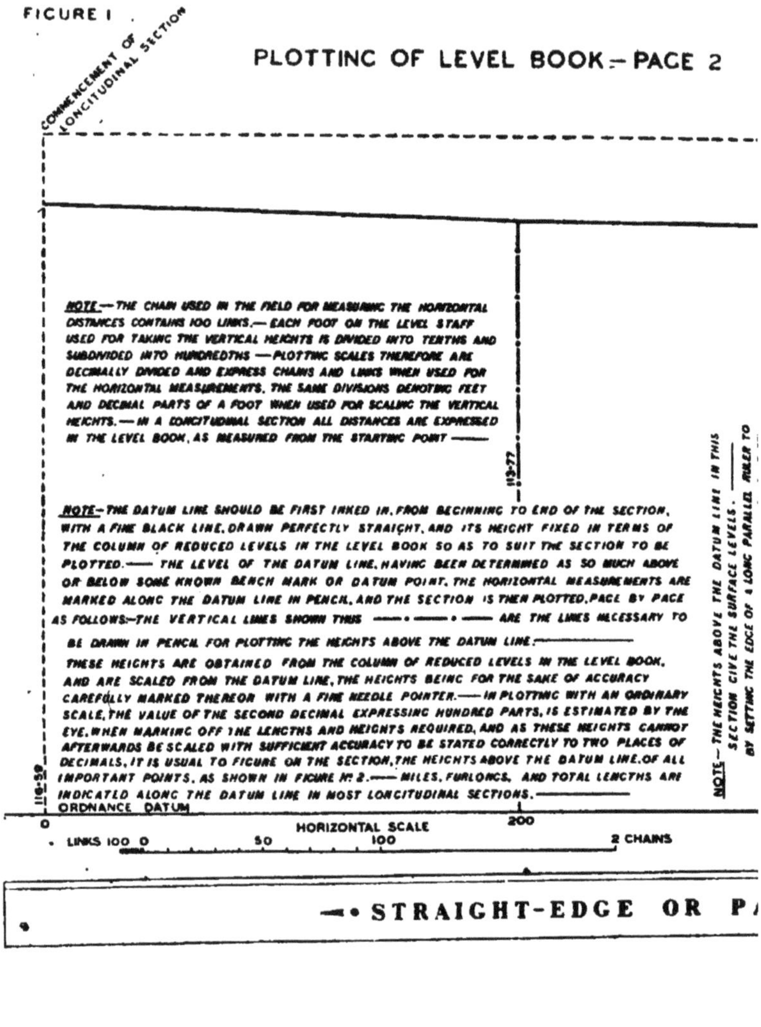

FIGURE I

COMMENCEMENT OF LONGITUDINAL SECTION

PLOTTING OF LEVEL BOOK — PAGE 2

NOTE — THE CHAIN USED IN THE FIELD FOR MEASURING THE HORIZONTAL DISTANCES CONTAINS 100 LINKS. — EACH FOOT ON THE LEVEL STAFF USED FOR TAKING THE VERTICAL HEIGHTS IS DIVIDED INTO TENTHS AND SUBDIVIDED INTO HUNDREDTHS. — PLOTTING SCALES THEREFORE ARE DECIMALLY DIVIDED AND EXPRESS CHAINS AND LINKS WHEN USED FOR THE HORIZONTAL MEASUREMENTS, THE SAME DIVISIONS DENOTING FEET AND DECIMAL PARTS OF A FOOT WHEN USED FOR SCALING THE VERTICAL HEIGHTS. — IN A LONGITUDINAL SECTION ALL DISTANCES ARE EXPRESSED IN THE LEVEL BOOK, AS MEASURED FROM THE STARTING POINT ————

112·77

NOTE — THE DATUM LINE SHOULD BE FIRST INKED IN, FROM BEGINNING TO END OF THE SECTION, WITH A FINE BLACK LINE, DRAWN PERFECTLY STRAIGHT, AND ITS HEIGHT FIXED IN TERMS OF THE COLUMN OF REDUCED LEVELS IN THE LEVEL BOOK SO AS TO SUIT THE SECTION TO BE PLOTTED. —— THE LEVEL OF THE DATUM LINE, HAVING BEEN DETERMINED AS SO MUCH ABOVE OR BELOW SOME KNOWN BENCH MARK OR DATUM POINT, THE HORIZONTAL MEASUREMENTS ARE MARKED ALONG THE DATUM LINE IN PENCIL, AND THE SECTION IS THEN PLOTTED, PACE BY PACE AS FOLLOWS: — THE VERTICAL LINES SHOWN THUS ——·——·—— ARE THE LINES NECESSARY TO BE DRAWN IN PENCIL FOR PLOTTING THE HEIGHTS ABOVE THE DATUM LINE. ————

THESE HEIGHTS ARE OBTAINED FROM THE COLUMN OF REDUCED LEVELS IN THE LEVEL BOOK, AND ARE SCALED FROM THE DATUM LINE, THE HEIGHTS BEING FOR THE SAKE OF ACCURACY CAREFULLY MARKED THEREON WITH A FINE NEEDLE POINTER. —— IN PLOTTING WITH AN ORDINARY SCALE, THE VALUE OF THE SECOND DECIMAL EXPRESSING HUNDRED PARTS, IS ESTIMATED BY THE EYE, WHEN MARKING OFF THE LENGTHS AND HEIGHTS REQUIRED, AND AS THESE HEIGHTS CANNOT AFTERWARDS BE SCALED WITH SUFFICIENT ACCURACY TO BE STATED CORRECTLY TO TWO PLACES OF DECIMALS, IT IS USUAL TO FIGURE ON THE SECTION, THE HEIGHTS ABOVE THE DATUM LINE, OF ALL IMPORTANT POINTS, AS SHOWN IN FIGURE Nº 2. —— MILES, FURLONGS, AND TOTAL LENGTHS ARE INDICATED ALONG THE DATUM LINE IN MOST LONGITUDINAL SECTIONS. ————

116·59

ORDNANCE DATUM

NOTE — THE HEIGHTS ABOVE THE DATUM LINE IN THIS SECTION GIVE THE SURFACE LEVELS. — RULER TO BY SETTING THE EDGE OF A LONG PARALLEL

HORIZONTAL SCALE
LINKS 100 0   50   100   200   2 CHAINS

——·— STRAIGHT-EDGE OR P

Land Surveying and Levelling, pp. 178, 179.

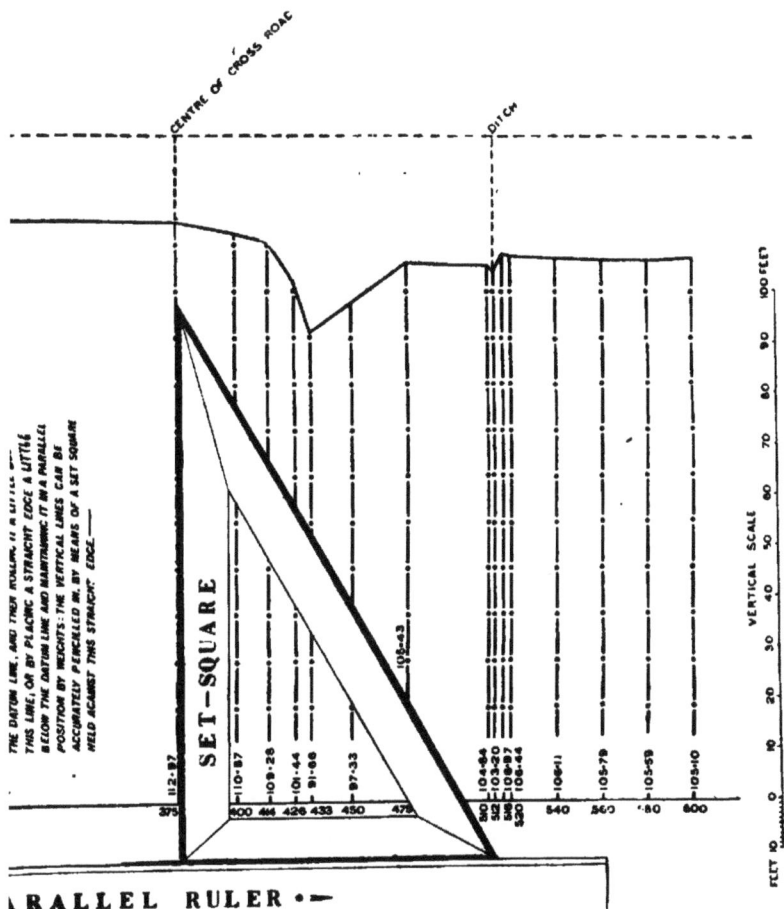

CHAPTER XVI.

LONGITUDINAL AND TRANSVERSE SECTIONS.

At least three assistants are required when taking a section, —two men at the chain and a man to hold the staff, and the more experienced and skilful a surveyor is in selecting the spot for setting up the instrument, the more rapidly will he get over the ground assigned to him. The staffman calls out the number of links upon the chain at which the staff is held, and the surveyor books the total number of links in the column headed "Lengths," and describes the various positions of the staff (when necessary) under the heading "Remarks." (See pages 173-176.) When the ground is of an undulating nature at close intervals, intermediate sights are required to be taken. These occur as shown in the form of Level Book above alluded to, between the back-sight and fore-sight at any single setting up of the level. The chain usually employed in the field for measuring the horizontal distances contains 100 links. Each foot upon the level staff used for taking the vertical heights is divided into tenths, and each tenth subdivided into hundredths.

In a longitudinal section (fig. 2, pages 182, 183) all distances are expressed in the Level Book as measured from the starting-point. The actual datum point need not be a point upon the line of section. In transverse sections (fig. 3, pages 186, 187), the distances are expressed as being so many links either upon the left hand or right hand of the centre line.

Plotting scales, as explained upon pages 92-94, are decimally divided, and express chains and links when used for horizontal measurements, the same divisions denoting feet and decimal parts of a foot when used for scaling the vertical heights (pages 178, 179). The vertical scale for a section is usually exaggerated

Land Surveying and Levelling, pp. 182, 183.

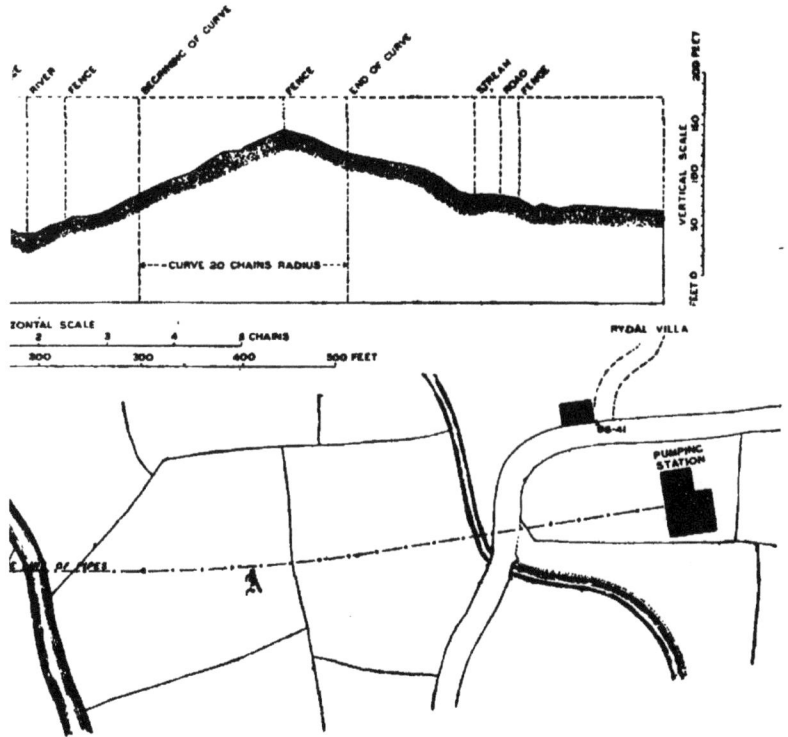

for the sake of accuracy, carefully marked thereon with a fine needle pointer held perpendicular to the drawing so as to puncture the paper as little as possible. A fine-pointed pencil is not accurate enough when the comparatively small scale of the section is considered. The 100th part upon the scale in plotting is estimated by the eye. Hence, so far as book work is concerned, there is no practical advantage in reading hundredths for an intermediate sight. The points so pricked off are joined by a fine black line ruled in ink, and this is best done if the surface line is inked in before plotting another page of the Level Book (page 174).

The rolling parallel ruler employed should be heavy, with handles to lift it, yet not too long for rapid use, and with its wheels deep enough to pass freely over the paper. A length of 18 inches proves satisfactory. If a long straight edge parallel to the datum line be used, it may be maintained in position by flat weights. The position of the datum point or bench mark is indicated upon the plan in fig. 2 near the end of the longitudinal section. Where the section is taken upon a curve, the radius of the curve (in this case 20 chains) is figured upon the longitudinal section. The tangent lines to this curve intersect at the peg A upon the plan (see fig. 2) off the line of the section. A horizontal line, as shown in fig. 2, is usually drawn above the section, parallel to the datum line. Dotted lines are drawn vertically from the line indicating the surface of the ground to this horizontal line at all points requiring description, and the description is generally written in at an angle of about 45° to this line, as shown. Where a peg is left in the ground upon the section at this point the vertical dotted line is carried down to the datum line, so as accurately to mark its position in a horizontal direction. (See pages 182, 183.)

Transverse sections (as shown in fig. 3, pages 186, 187) have their position marked upon the longitudinal section as well as upon the plan. The height above the datum line is figured upon each transverse section, and also upon the longitudinal section, where the cross section occurs. They are taken for a distance of either one or two chains as required upon each side of the centre line, and in order to continue the levelling between the bench marks, they are levelled alternately thus :—C. S. No. 1 is taken from right to left, C. S. No. 2 from left to right, C. S. No. 3 from right to left, viewed in the direction of the chaining of the longitudinal section, as indicated by the arrows upon the plan. This continuous method is found in practice to be the most satisfactory way of connecting transverse sections with a longitudinal section, as the latter can be more expeditiously proceeded with, where there are numerous cross sections required, than if the cross sections are taken all in the same direction. They are, however, all plotted in a direction to read from left to right. Sometimes, at the outer end of a transverse section upon falling ground, the level staff may appear to be a little short, in which case, rather than change the position for a sight, which, though needed to complete the section, may not at this distance from the centre line seriously affect the future work, a surveyor will instruct his staff holder to raise the staff one foot or two feet, which another man could measure against a ranging-rod, and thus obtain a view of the sub-divisions of the Level staff upon the line of collimation.

In the case of a river section, a cord or wire may be stretched between the two opposite banks, and horizontal distances subsequently marked thereon, after it has assumed its final sagging, by attaching short pieces of tape at every 5 or 10 feet, but it will be found more expeditious to employ a rope on floats pulled tight, and to measure the horizontal distances upon this rope. Having taken the water level very accurately, and noted the hour of the day in so doing, in case its level may be affected by tidal action, soundings are taken at the points indicated, from which a section can be plotted, giving the transverse bed of the river. The longitudinal section of the bed of the river could be subsequently made up from several cross sections, accompanied by an accurate plan.

When levelling in connexion with proposed works likely to be influenced by flood levels, ordinary or extraordinary, it is necessary to gain all the information that can be acquired as to the different levels of brooks, streams and

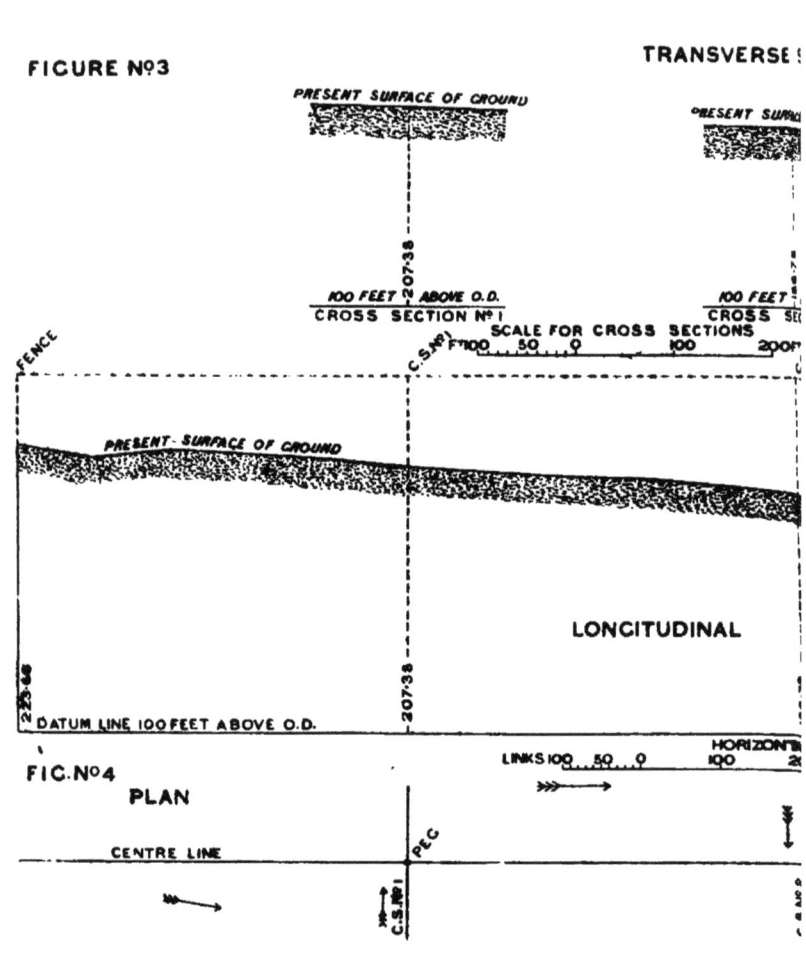

Land Surveying and Levelling, pp. 186, 187.

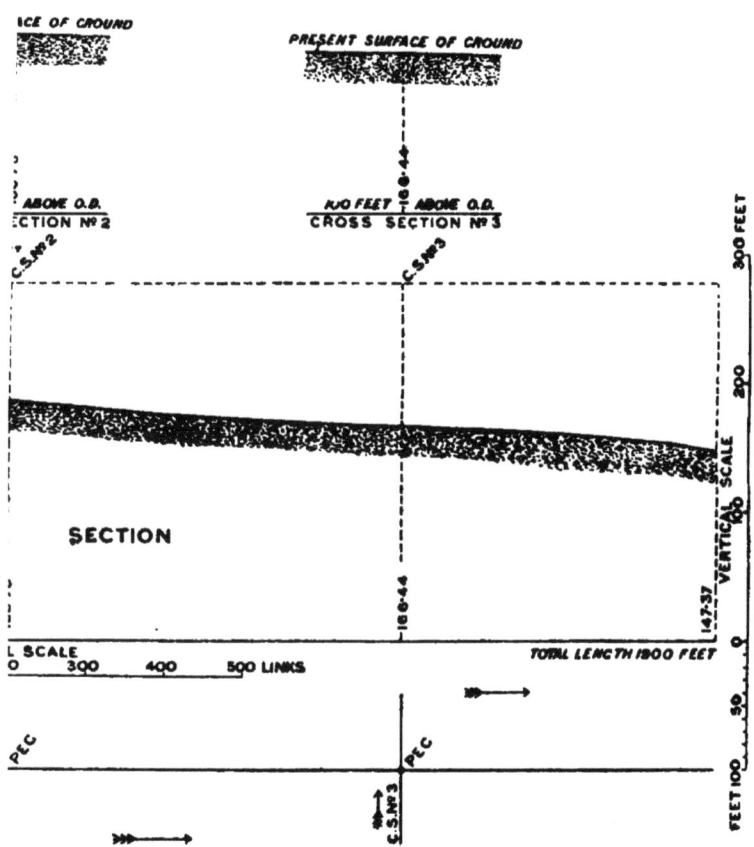

SECTIONS

rivers crossed, and to enter all the data so learnt in the page of the level book headed " Remarks."

In order to approximately gauge the volume of water flowing in the channel of a stream, it is necessary to set up an artificial dam, which is usually constructed of stout planking, in which a rectangular notch is cut; for a stream, this notch will rarely exceed a foot or a foot and a half in depth; the edges of this notch should be chamfered at an angle of $45°$, and be neatly lined at the crest with iron. This overfall should be strongly built, and with close joints; it must be solidly set up, so as to resist without bending, swerving, or leaking, the body of water which will accumulate behind it; it must be set up truly level, with the sharp crest facing the current of the stream, also sufficiently high out of water, so that the sill will not be "drowned," by floods, but at the same time not so much as regards streams more particularly, as to cause the adjacent fields to be inundated by the first rainfalls, for, besides other consequences, there would then be an end of the stream gauging for a time.

It is often very important to ascertain the quantities arising from flood waters. Before setting up the overfall, a stout post should be firmly driven into the bed of the stream some seven or eight feet above the site of the overfall, and as soon as this latter has been well and soundly got in, the post should be cut down exactly to the level of the sill of the overfall; this should be adjusted with a spirit level; a sharp-edged rule, decimally divided (in feet), should then be fixed to the post with the zero level with the edge of the sill on the overfall; by means of this rule the depths of water flowing over the notched board will have to be measured. When from heavy rains the stream swells so that the waters approach the overfall with an appreciable velocity, it will have to be taken into account. With regard to the length of the sill of the overfall, it should not be much less than the width of the channel in which it is set up, or it will cause a wire drawing of the stream. In a wide river, the system explained under the head of Marine Surveying will probably have to be adopted. (See pages 226-233.)

LONGITUDINAL AND TRANSVERSE SECTIONS. 189

In the process of taking levels, an experienced surveyor will, in fine weather, endeavour to sight as much as possible towards the west in the forenoon, and towards the east in the afternoon, so as to have the sun as much as possible upon the level staff. He will stand with his back to the sun, as it is better to have the sun upon the staff than upon the object glass of the instrument.

It sometimes happens in the course of running a section, that the surveyor comes upon a very abrupt difference in the level of the ground, too great to connect with the level-staff, such as a vertical rise or fall of 20 feet, more or less. The best plan to adopt in such a case, is to level round without chaining more than the line of section requires, but if the ground does not admit of such connexion of levels within a reasonable distance, then a break must be made in the series of levels and the exact height checked by means of a tape or a measuring rod. The levels of both the upper and lower points should be arrived at by levelling from distinct bench marks, in order to determine the correct difference of level accurately. As with the use of an ordinary plotting scale, the value of the second decimal expressing hundred parts can not be scaled with sufficient accuracy, especially with liability to shrinkage in the paper upon which the section is drawn, it is usual to figure upon the section the heights above the datum line of all important points. It is also usual to add the total length in miles, furlongs and chains at the end of the line of section. Horizontal scales are best drawn horizontally. Vertical scales are best drawn vertically on the paper. (See pages 182, 184.)

The late Principal of the Crystal Palace Company's School of Practical Engineering at Sydenham, when addressing his students in 1876, gave them the following valuable hints: "Always keep your book clear and distinct; do not be afraid of entering too many particulars; take plenty of bench marks when you are levelling, for future reference; keep the glasses of your instruments clean, but do not fall into the error of one of my young friends who, being on a survey, telegraphed to me one morning to this effect, 'I have thoroughly cleaned out the level from end to end with my pocket handkerchief, and for the life of me I cannot see any cross hairs.'" The principal then went on to say that "a good surveyor will practise the judging of distance, and will often test his ordinary paces when walking, so that he can obtain the length of a line pretty accurately without the aid of a tape or chain. It often happens that he has to cross forbidden ground over which he cannot chain a line, and this plan will be found most useful. I was once thus stepping the distance across a large field, the property of an irascible farmer, and had proceeded nearly two-thirds of the way when he confronted me with a big stick, saying, 'Hi, you. What business have you on my property? You're a spoiling of these turmits.' 'My good friend, I am doing no harm.' 'I'm no friend of yours; I am your henemy. Come, get out of this, can't yer?' 'Well, then, my good enemy, you see I have already got two-thirds of the way across this field, and it will do the turmits, as you call them, less damage if I go off them in this direction.' 'Wall, may be, it wull; but I'll see you safe off afore I goes away.' 'Thank you. Good afternoon. 196, 197, 198,' and so on until the line was complete. Sometimes, however, an opposing landowner knows what he is about much better than this simple farmer, and puts the surveyor to considerable inconvenience. Imagine the disgusting case of a friend of mine, who in taking a line of levels was constantly confronted by a couple of rustics holding up a large piece of rick-cloth stretched between two poles. Who could object? They were at liberty to stand where they liked, and hold what they liked, and the old farmer stood by and chuckled derisively At length, with a bow, the surveyor retired. The next morning the farmer and he met again some distance ahead. 'Ah, old fellah!' said Hodge, 'I did yer last night. I know a thing or two.' 'Yes,' quietly remarked my friend, 'but you don't know everything. I was back after dark and completed my line of levels with the aid of a lantern.' It is not an easy thing to take levels with the aid of an artificial light, but it may be done, and it is useful to know that, with a little care, it can be effected."

CHAPTER XVII.

CONTOURS.

A CONTOUR plan enables the surveyor to judge the undulatory nature of any ground, and in setting out work to run lines approximately of equal level or of uniform gradient. (See pages 194, 195.)

They are plotted upon a plan by lines representing the imaginary intersection of horizontal planes with the surface of uneven ground, and form the outline of a horizontal section or the outline of any portion of the hill made by still water rising to that height. The figures written upon the chain dotted lines in the accompanying plan, represent vertical intervals of five feet. The plan is prepared in the following manner. Suppose the altitude of the contour required to be 120 ft. above a given datum (see page 177), and that the nearest reliable bench mark is known to be at a level of 137·86 ft. above the same datum, there is a difference of 17·86 ft. between the level of the bench mark and the required contour. The instrument is set up in a convenient position and adjusted, and if, for example, the staff reads 2·61, this is booked as a back sight and shows the line of collimation to be at a level of (2·61 plus 137·86) 140·47 ft. above the given datum. Now the contour being 20·47 ft. below this, it is evident that the line of collimation must be lowered in position before a level staff, 14 ft. or 16 ft. long, could be read by the instrument when set up level. A fore sight for the purpose of connecting the levels is taken in such a position that the staff is read near the top, say at 12·85, and the level is re-set in a new position to read near the foot of the staff, say 1·83. Reducing the book we find the level of the point upon which we changed the level of the line of collimation to be 127·62. The position of this point is not recorded in the level book, as it is not required to be indicated upon the plan, but we have by this means fallen from a known level of 137·86 to a known level of 127·62, and have now only to fall 7·62 ft. lower to find the position of a point upon the required contour. Adding 7·62 to 1·83 we obtain 9·45 as the necessary reading upon the staff to give the height required, and this figure is then booked for reference in the intermediate column. If the staff reads less than 9·40 the staffman is told to go to lower ground, whereas if the staff reads more than 9·50, the staffman is told to come to higher ground. Except where extreme precision is required, the divisions indicating hundreds may be disregarded in reading the staff, but it is well to observe them in the Level Book for sake of accuracy in connecting the levels. So far as points of equal level are concerned, a reflecting level would serve to fix them approximately, but the most accurate work is secured by the aid of a Dumpy level as above described. Signals will have to be agreed upon by which the surveyor can communicate with the staffman when they are too far apart to be within hearing of one another. The position of the contour points in the field is fixed during the process of levelling by stout laths pointed at the ends, so as to be left temporarily in the ground at suitable distances apart. The number of laths required will entirely depend upon the intricacy of the ground and the size of the features. Laths are placed in positions determined by levelling at such points as will best define the shape of the contour, and these laths are afterwards surveyed in a similar manner to points in a fence. It is advisable only to level and lath out as much as can be surveyed before leaving the ground for the day.

Should the contour pass round the shoulder of a hill, so that the staffman in approaching the next point is out of sight of the surveyor at the instrument, or should the distance become too great for the staff to be distinctly read through the telescope, the instrument must be moved to a fresh position, and the levelling connected. Suppose, after fixing the contour laths marked A, the surveyor decides to shift the instrument, and directs the staffman, for the purpose of securing accuracy in the change, to hold the staff upon the changing peg or footplate. If the foresight reads

9·31, the height of the changing point will be 0·14 above the contour level. Hence, when the instrument is re-set up in the new position, if the back sight reads 6·78, the intermediate sight for fixing the contour laths marked B in the diagram must be 6·92, so that the reduced level may again become 120·00. In like manner when again changing the position of the instrument between the contour laths marked B and those marked C, if the changing point be 0·06 below the contour level, the intermediate sight then taken must be 0·06 above the back sight 8·29, in order to arrive at the correct reduced level for continuing the line of contour. It will thus be seen that *by reducing the Level Book in the field as the work proceeds*, the contour points upon lines of equal level can be most readily determined.

When an extensive contour plan is required to be completed in a short time, one set of men may be engaged to fix the contour laths by a process of levelling, and another set of men to survey them, but when more than one line of contours is set out upon the ground, it is necessary to carefully distinguish between the laths indicating the higher and lower contours, and for this purpose three distinct sets, painted respectively either red, white, and blue, may be employed to be used in turn, one set after the other for the individual contours.

In practice an intelligent man may very soon be made to get into the way of judging the correct distance or height to traverse; which is a considerable saving of time to the surveyor.

In contouring a small enclosure the theodolite sometimes is set up level in such a position that by clamping the vertical arc the required contour can be traced completely round; but when the area is of some extent the process illustrated in the diagram by means of a Dumpy level and staff is by far the most satisfactory. In order to prove the accuracy of the day's work, it is well to leave off levelling upon a bench mark as when taking an ordinary section.

Contour lines may be required to be traced round isolated features of ground for setting out water channels, railways, or for other purposes. In contouring a mountainous country the watershed of the small ranges which project

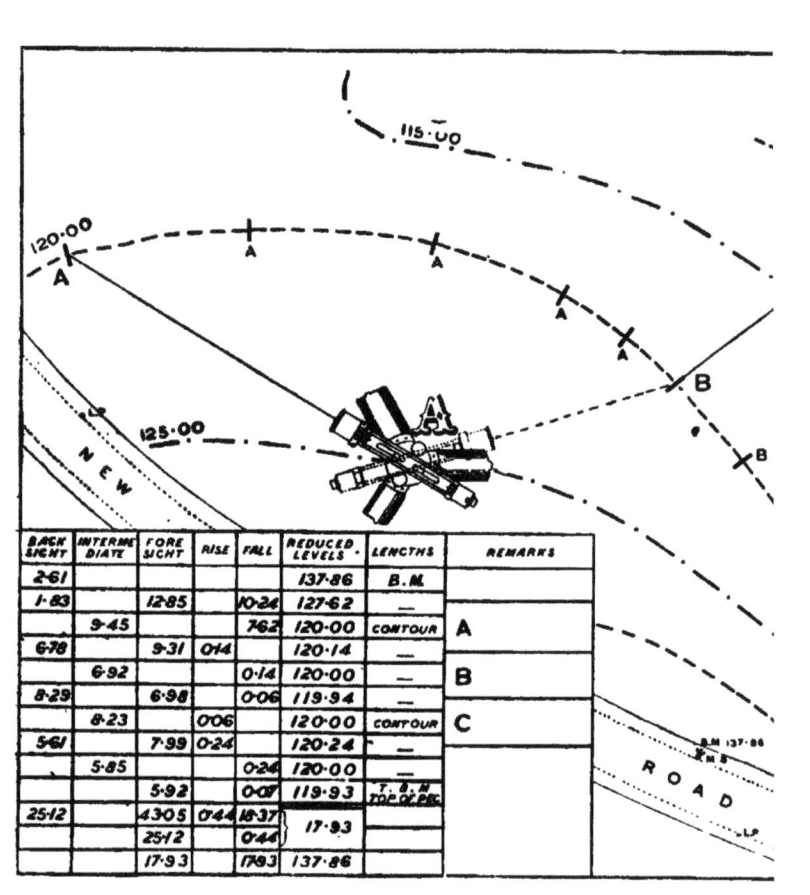

BACK SIGHT	INTERME DIATE	FORE SIGHT	RISE	FALL	REDUCED LEVELS	LENGTHS	REMARKS
2·61					137·86	B.M.	
1·83		12·85		10·24	127·62	—	
	9·45			7·62	120·00	CONTOUR	A
6·78		9·31	0·14		120·14	—	
	6·92			0·14	120·00		B
8·29		6·98		0·06	119·94	—	
	8·23		0·06		120·00	CONTOUR	C
5·61		7·99	0·24		120·24	—	
	5·85			0·24	120·00		
		5·92		0·07	119·93	T.B.M. TOP OF PEG	
25·12		43·05	0·44	18·37			
		25·12		0·44	17·93		
		17·93		17·93	137·86		

Land Surveying and Levelling, pp. 194, 195.

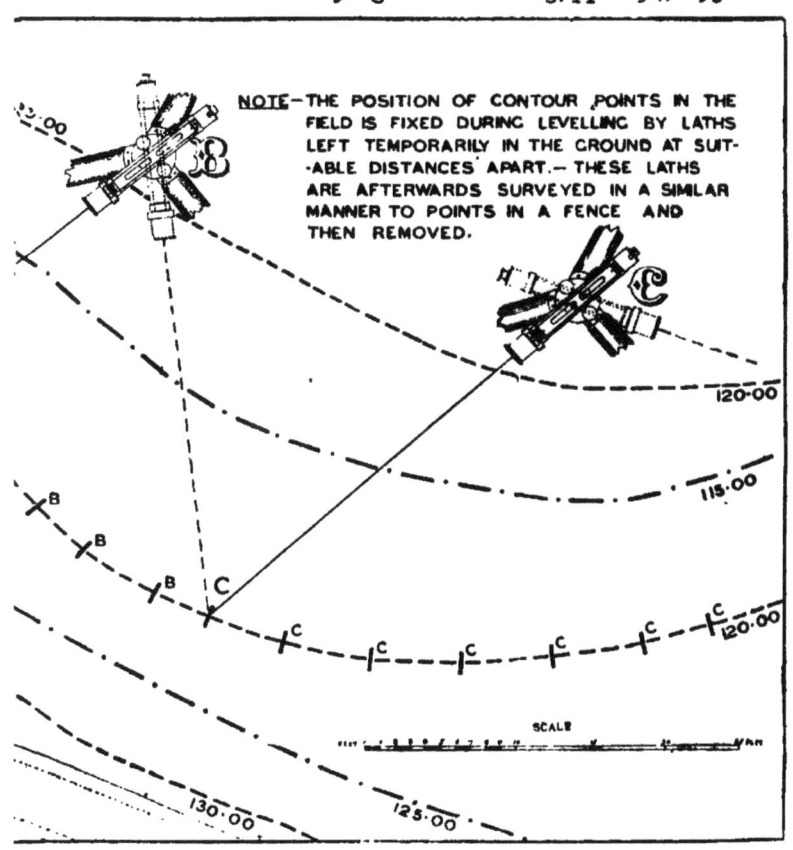

laterally or at right angles from the larger ranges, will generally be found to have the most gradual and regular slope. Hence the datum pegs for checking the contour levels in the process of setting out can best be fixed at such points. When running a line similar to a contour for setting out a gradient, the assistance of a chain is required to arrive approximately at relative distance. Two men are, therefore, engaged to chain continuously over the line marked out by the laths, following by eye the irregular course so indicated as near as possible. Suppose that the object of the contour is to subsequently set out a gradient of 1 in 1,000. The follower at the chain reports to the surveyor at the insertion of every fifth arrow, indicating 500 links with the use of a foot chain, or at every $7\frac{1}{2}$ chains (497 feet) with the use of a Gunter's chain. The surveyor then either reduces or increases the reading to be taken upon the staff by 0·50, according as he is ascending or descending a gradient. It is not necessary to alter the reading for determining the position of these points more frequently than at intervals of, say, half a foot, because the line afterwards ranged out for adoption when marked out either as a straight line or set out as a regular curve, is not likely to approach nearer to the line of levels so found than six inches when taken as a longitudinal section.

As regards the origin of contour lines, Mr. J. Butler Williams, in his treatise on Practical Geodesy, states that "the idea of employing horizontal lines for the representation of forms suggested itself as early as 1742 to Philippe Buache, when observing the horizontal marks left on the land by the gradual subsidence of the waters of an inundation. But he only aimed at tracing them as lines of equal soundings on hydrographical charts. Thirty-three years later, Ducarla proposed to adapt imaginary lines, following the same law, to the representation of the features of ground; and in 1782, Dupain-Triel in arranging the method into a system, recommended that the lines should generally be vertically equidistant. However, owing to the want of precise data and accurate outline maps, the principle was not fully carried out until of late years, when these vertically equidistant horizontal lines, which we shall call normal contours, came to be determined in the field by means of the theodolite, spirit level, or other instrument. These normal contours have been adopted in the French cadastre since 1818."

The Ordnance Surveyors employed a 5-in. theodolite for contouring with the aid of a contouring staff 8 ft. long figured upon one face only, and provided with a sliding vane which was fixed at the required height by a clamp at the back of the staff. The visibility of points upon a site to

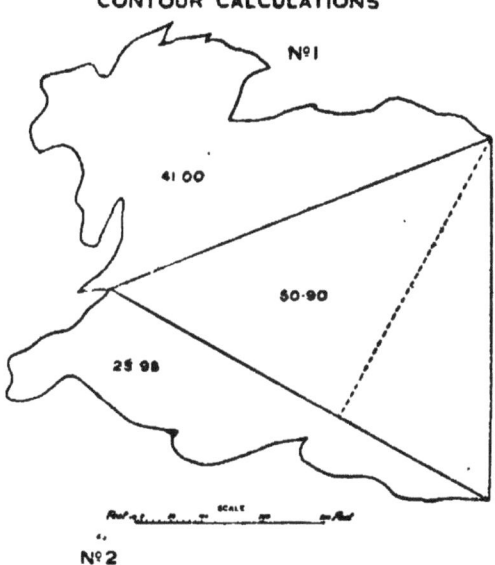

be seen from one another is sometimes an important question. This depends mainly upon the general convexity or concavity of the intervening ground, and may best be determined by a longitudinal section taken between the points under review, but in the absence of such section, it may be judged by a glance at a plan showing the intervening contours, bearing in mind that when the contour lines which were run at the higher level appear closer together than those which were run at the lower level, the

198   LAND SURVEYING AND LEVELLING.

ground must be concave, and that when the contour lines at the higher level appear further apart than the contour lines at the lower level, the ground must be convex.

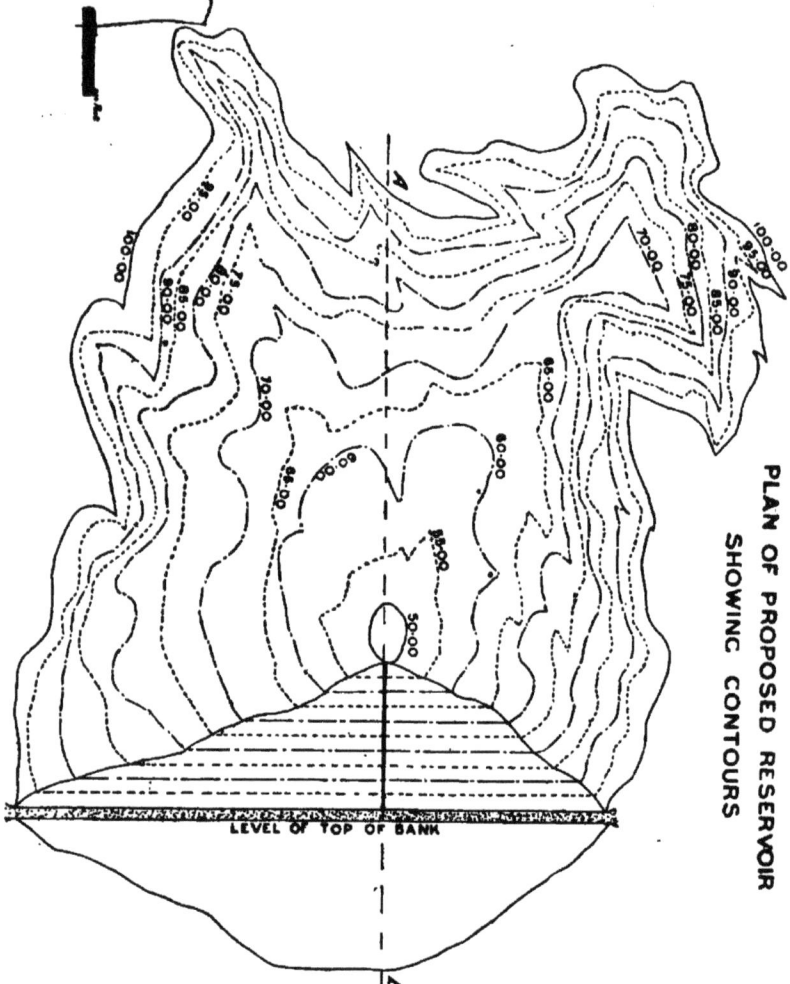

The plan of the proposed reservoir shows contours at heights having an interval of 5 ft. between them. The altitudes are supposed to range between 50 and 100 ft. Watersheds are indicated by a salient or outward bend in a contour line; watercourses by a re-entering bend in the contour line. The vertical interval at which the contours are to be levelled having been decided, a line of section is taken (sometimes called "the initial line"), upon which pegs, as temporary bench marks, are fixed near the level of each contour, and in running the required contours the instrument is set up so that the telescope, when levelled, may intersect the levelling staff which is held upon the initial point, or top of each contour peg, in order to arrive at the reduced level to be read upon the staff for tracing the required contour, as explained above.

The areas of the contours can be best arrived at with the use of a planimeter, as described upon pages 155, 156. In carrying the tracer round, care must be taken to observe whether the disc M has performed an entire revolution, in which case 10 must be added to the unit in the last reading taken by the instrument. The diagram indicates the method of calculating the area enclosed by the top bank contour. In this case the enclosure being large it is divided into three parts, and one side being straight, a triangle is formed in the centre, the area of which can be most accurately calculated in the ordinary way by multiplying the base by half the height, as it would be waste of time to go round the boundary of any regular geometrical form with the planimeter. The two adjoining enclosures to the triangle being irregular in outline are more accurately determined by the planimeter than by drawing "give-and-take lines" in the usual way. (See page 197.)

The subjoined table gives a systematic method of drawing up the statement of capacities. (See page 200.)

There are two principal modes of expressing the nature of hill and valley on a plan. The vertical system is largely employed in France and Prussia, and also in England when the scale is very small and the maps are to be engraved; and in this method it is assumed that the lines represent the course which the rills of water would take in descending from the water-sheds to the lower levels. It may be used simply for the expression of ground when the relative

LAND SURVEYING AND LEVELLING.

thickness or distance of the lines apart have no especial meaning beyond depicting the folds and irregularities of the ground; or the lines and distances may be constructed according to a scale of shade, as in German military maps, where the relative thickness denotes the *angle* of the slope

CALCULATION OF CAPACITIES.

PROPOSED RESERVOIR.

Reduced level.	Sq. ins. area on paper.	Area in sq. feet.	Mean Areas.	Depth.	Cubic feet.	Content in cubic feet up to level.	Capacity in gallons up to level.
50·00	0·46	1656					
			6912	× 5	34560	34560	55·00 / 216000
55·00	3·38	12168					
			28638	× 5	143190	177750	60·00 / 1110938
60·00	12·53	45108					
			60786	× 5	303930	481680	65·00 / 3010500
65·00	21·24	76464					
			100188	× 5	500940	982620	70·00 / 6141375
70·00	34·42	123912					
			150228	× 5	751140	1733760	75·00 / 10836000
75·00	49·04	176544					
			199003	× 5	995040	2728800	80·00 / 17055000
80·00	61·52	221472					
			244872	× 5	1224360	3953160	85·00 / 24707250
85·00	74·52	268272					
			291528	× 5	1457640	5410800	90·00 / 33817500
90.00	87·44	314784					
			339246	× 5	1696230	7107030	95·00 / 44418937
95·00	111·03	363708					

Area of Top Bank Level 9·58 *acres.*

as well as its form. Thus Lehman's scale of shade commences with fine lines at 5°, and proceeds in increasing thicknesses of line, every 5°, up to 45°, which being practically impassable is represented as composed of very thick black lines extremely close to one another.

The horizontal method may be considered as but an extension of the principle of contouring, as the space between the chief contours is filled in with horizontal lines, preserving a rough parallelism to them. These hachures are not, however, continuous, as this would tend to make the drawing too smooth and even, and destroy the irregularity of work produced by drawing them in sets or groups of various lengths, which, moreover, tends to produce a nearer resemblance to the roughness of the actual surface.

Lieut.-Colonel William Paterson, Professor at the Royal Military College, Sandhurst, in his "Notes on Military Surveying and Reconnaissance," thus describes the different theories of expressing ground upon paper: "There are three principal and distinct methods of expressing hills upon paper. The first method is to suppose a sheet of light in parallel rays to fall upon ground from directly above it. Wherever the ground is level, the appearance will be white; when it slopes, the rays falling obliquely on the surface will give a dark appearance; and as it becomes more precipitous, it will of course be still darker. This is the theory of the style of drawing which is done with a brush and successive shades of indian ink or other tint. In the second method, a volume of water is supposed to fall on a hill. Wherever the slopes are gentle, the water will run off in small thin rills, diverging as they descend; in the abrupt slopes, it will rush down in large broad streams. This is the theory of the vertical system; only to avoid stiffness, the lines are broken into a series of successive strokes, not following one another down the hills. In the third method, the hills may be imagined as inundated with water, and that every time the water falls, a certain fixed number of feet, a mark is run round the surface of the ground at the upper edge of the water. According to the steepness of the falls, the lines, to an eye placed immediately above them, will appear close or far apart. This is called the contour or horizontal system, and is applied in field sketching; the strokes being broken instead of being carried continuously round, and the effect of the shading increased by making them thicker, as well as closer on the steeper slopes, and lighter, longer, and wider apart on the more gentle declivities." (See also remarks upon pages 120-122.)

CHAPTER XVIII.

SETTING OUT CURVES.

CURVES may be set out by means of deflection angles. As shown in the diagram illustrating Case 1 (pages 206, 207), the tangential angle is half the deflection angle and half the central angle. Though it is not necessary to set out a curve geometrically for an ordinary road, yet it is always a neat method to join two straight portions of a bending road with a proper curve, while in railway work it is absolutely essential. In the case of an ordinary road it facilitates the setting out of a line of frontage for building. In railway work it is the English practice to always introduce a length of straight line wherever possible between two curves whether similarly inclined or reversed. Usually a length of at least two chains should intervene. The Parliamentary regulations provide for curves of more than half a mile radius to be sharpened if required in construction to half a mile radius, while curves of less radius may not be sharpened. As a limit of curve, 15 chains or sometimes a two-mile radius is adopted for main lines. (See page 265.)

The angle between a tangent line and a chord being proved by the consideration of Case 1 to be equal to half the angle at the centre of the circle subtended by any given length of chord, we are enabled to set out a curve as shown in this diagram, the length of the successive chords and chord lines produced being usually measured with a chain and the offsets with a tape. The unit angle, or angle in which the chord is equal to the radius, is seen to be 57·295 degrees at the centre of the circle, giving a tangential angle with the chord equal to 28·648 degrees, or 1,719 minutes nearly. This unit tangential angle serves as a *constant*, which enables us to set out various points in a

SETTING OUT CURVES CASE I –

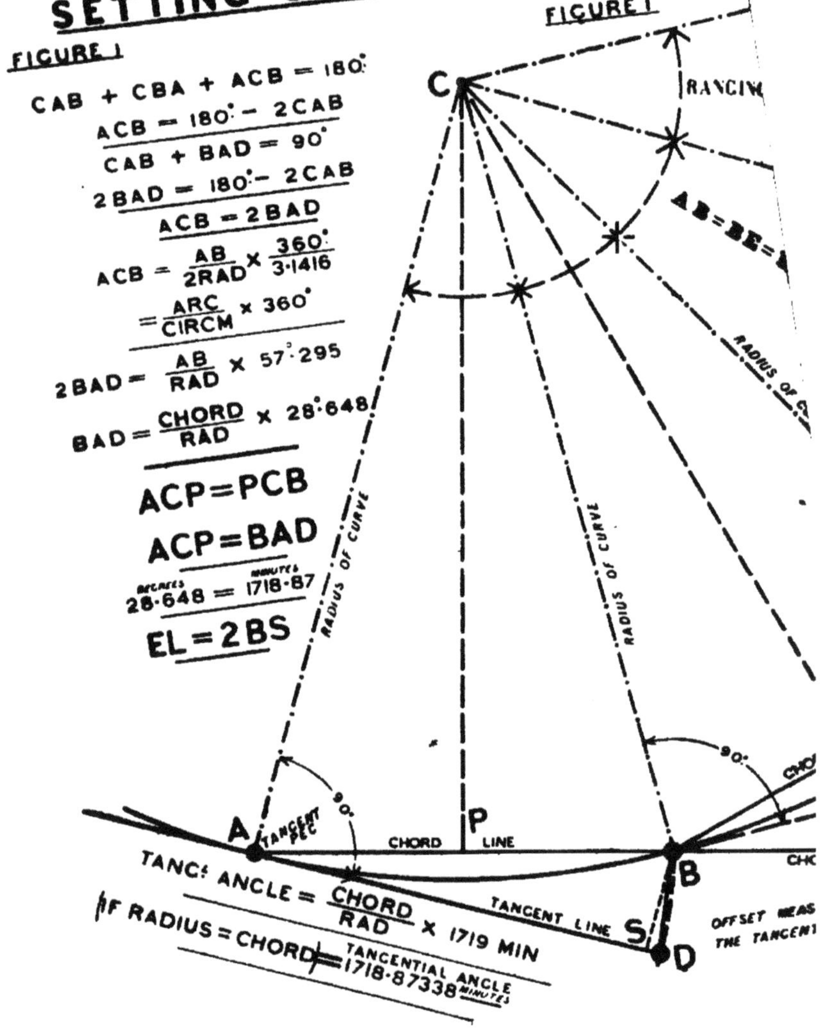
FIGURE I
FIGURE I
CAB + CBA + ACB = 180°
ACB = 180° – 2CAB
CAB + BAD = 90°
2BAD = 180° – 2CAB
ACB = 2BAD
ACB = AB/2RAD × 360°/3·1416
= ARC/CIRCM × 360°
2BAD = AB/RAD × 57°·295
BAD = CHORD/RAD × 28°·648
ACP = PCB
ACP = BAD
28·648 DEGREES = 1718·87 MINUTES
EL = 2BS
C
RANGIN
AB=BE=
RADIUS OF C
RADIUS OF CURVE
RADIUS OF CURVE
90°
90°
A
TANGENT PEG
CHORD
P
LINE
B
CHO
TANGENT LINE
S
D
OFFSET MEAS
THE TANGENT
TANG^t ANGLE = CHORD/RAD × 1719 MIN
IF RADIUS = CHORD = TANGENTIAL ANGLE 1718·87338 MINUTES

Land Surveying and Levelling, pp. 206, 207.

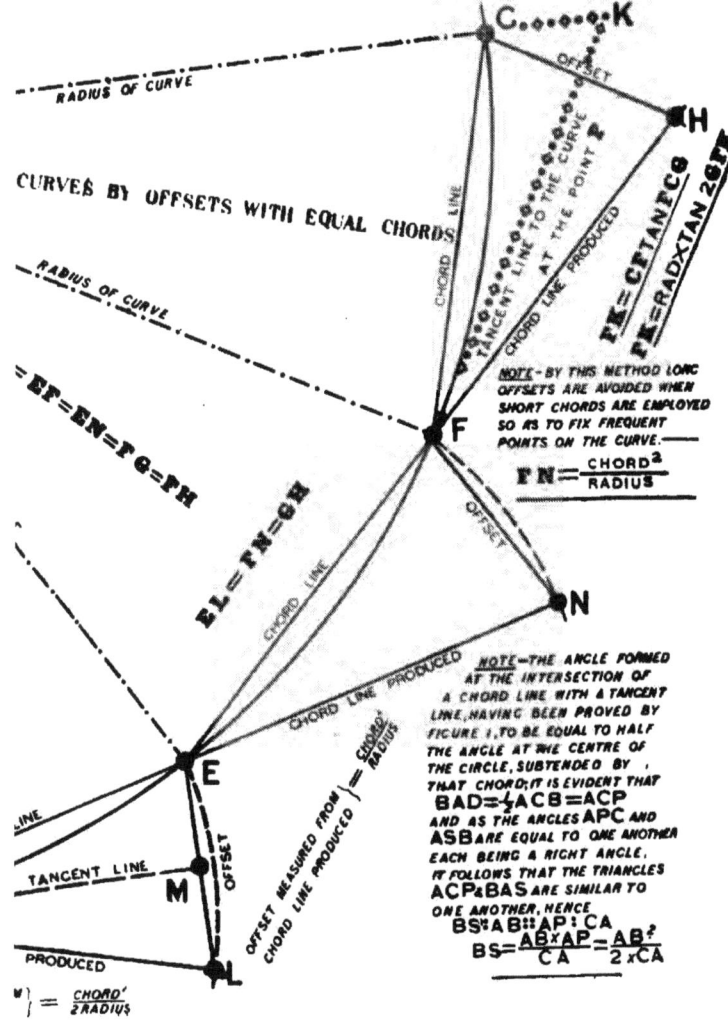

have to refer only state, when single minutes are required, the values of the sines and cosines of angles; thus,

$$\text{Cot } 62°\ 27' = \frac{\cos 62°\ 27'}{\sin 62°\ 27'} = \frac{\cdot 4625225}{\cdot 8866075}$$

The reduction of this vulgar fraction, which will be found equal to ·5216767, will show the value of the application of logarithms.

The above approximate result, ·521719, is seen to be correct for three places of decimals, but for application to the plotting of long base-lines the more accurate process must be applied.

To determine intermediate points in the curves between F and H, and between G and H (fig. 2), we can apply the formula given in Case 1 for the measurement of offsets from the tangent lines F B and G B, as it will be observed in the diagram which illustrates Case 1, that so long as the tangential angle B A D remains comparatively small, the length calculated for B S may be measured from D at right angles to the tangent line without appreciable error, to determine a point in the curve. Hence, substituting the term "distance" measured along the tangent line for the term "chord" in the equation (Case 1), we obtain for Cases 2 and 3 the formula,—

$$\text{Offset} = \frac{\text{distance}^2}{2\ \text{radius}}$$

If, therefore, the square of the number of links in the distance from the tangent peg (fig. 2) measured along the tangent line to the point at which the offset is desired to be taken be divided by the length expressed in links of the diameter of the circle, part of the circumference of which forms the required curve, the result will give the approximate number of links in the offset, the length of which is to be measured with the tape or measuring rod in a perpendicular direction to the tangent line. The formula is based upon an assumed length of radius, and the points upon the curve fixed thereby can be marked in the field by laths, pointed at the ends, so as to be easily pressed into the ground. (See pages 210, 211.)

Where great accuracy is not required, the application of

SETTING OUT CURVES. 209

the above formula derived from Case 1 (pages 206, 207) may be adopted, by the method shown in Case 3, for setting out offsets at right angles to a tangent line (pages 210, 211). The first offset nearest the junction of the curve with the tangent line is thus easily calculated, being equal to the length measured along the tangent line, divided by twice the radius of the curve, and if the remaining offsets be set out at the same distance apart, measured along the tangent line, the succeeding offsets, Nos. 2, 3, 4, and 5, will be respectively equal in proportion to the square of these numbers, or 4, 9, 16, and 25 times the length of the first offset. As the middle point in the curve is approached, the tangential

FIGURE 3

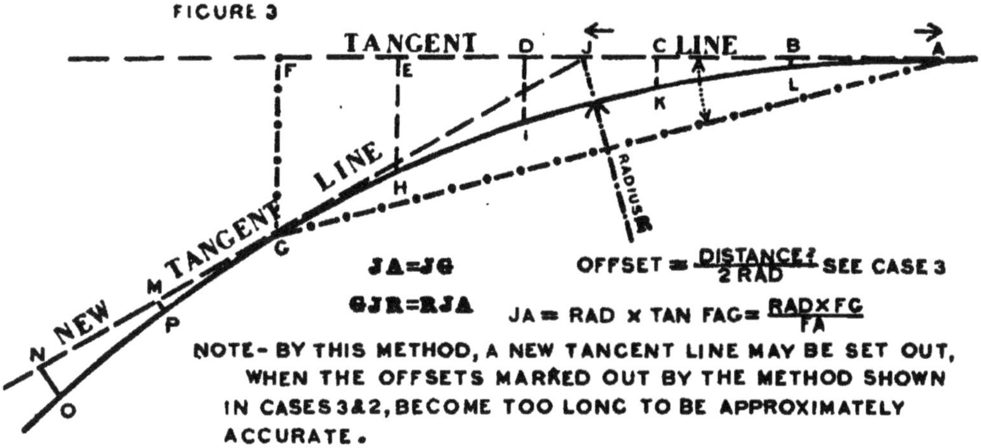

$JA = JG$

$GJR = RJA$ $JA = RAD \times TAN\ FAG = \dfrac{RAD \times FG}{FA}$

$OFFSET = \dfrac{DISTANCE^2}{2\ RAD}$ SEE CASE 3

NOTE- BY THIS METHOD, A NEW TANGENT LINE MAY BE SET OUT, WHEN THE OFFSETS MARKED OUT BY THE METHOD SHOWN IN CASES 3 & 2, BECOME TOO LONG TO BE APPROXIMATELY ACCURATE.

angle between the chord and the tangent line increases, and the amount of error in the length of the offset so calculated will also be increased. This error can, however, be kept small by setting out a new tangent line as shown above in fig. 3.

When the angle formed at B is smaller than that shown in Case 4, fig. 1 (pages 214, 215), it will probably be found a more convenient mode for setting out the curve to adopt one of the following methods:—(1) In the case of a radius not greater than one chain, to divert one or both or a portion of one of the lines A B and B C, making them parallel to one another.

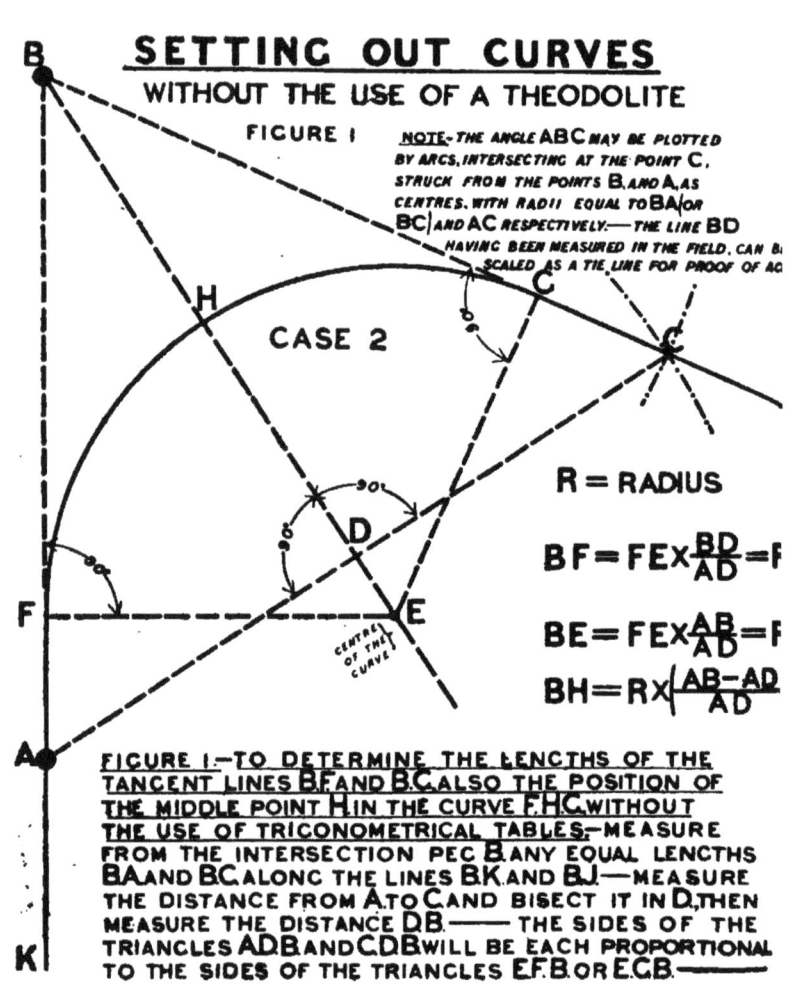

Land Surveying and Levelling, pp. 210, 211.

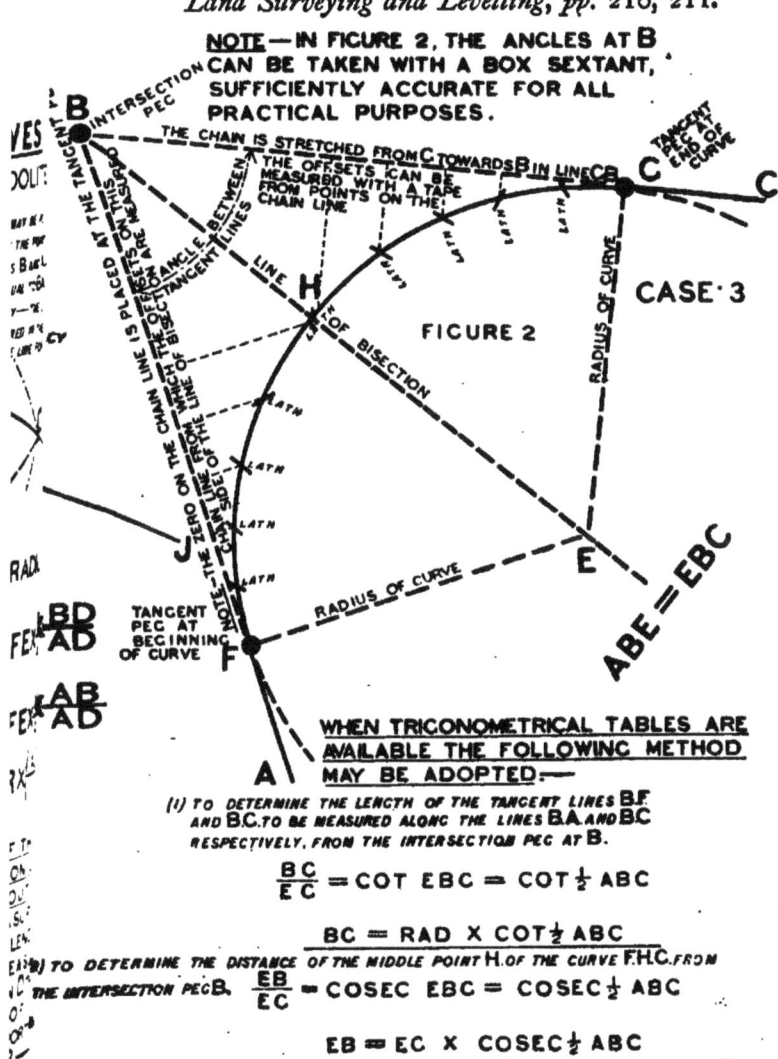

NOTE — IN FIGURE 2, THE ANGLES AT B CAN BE TAKEN WITH A BOX SEXTANT, SUFFICIENTLY ACCURATE FOR ALL PRACTICAL PURPOSES.

FIGURE 2

CASE 3

ABE = EBC

WHEN TRIGONOMETRICAL TABLES ARE AVAILABLE THE FOLLOWING METHOD MAY BE ADOPTED:—

(1) TO DETERMINE THE LENGTH OF THE TANGENT LINES B.F. AND B.C. TO BE MEASURED ALONG THE LINES B.A. AND B.C. RESPECTIVELY, FROM THE INTERSECTION PEG AT B.

$$\frac{BC}{EC} = \text{COT } EBC = \text{COT } \tfrac{1}{2} ABC$$

$$BC = RAD \times \text{COT } \tfrac{1}{2} ABC$$

(2) TO DETERMINE THE DISTANCE OF THE MIDDLE POINT H. OF THE CURVE F.H.C. FROM THE INTERSECTION PEG B.

$$\frac{EB}{EC} = \text{COSEC } EBC = \text{COSEC } \tfrac{1}{2} ABC$$

$$EB = EC \times \text{COSEC } \tfrac{1}{2} ABC$$

$$BH = EB - RAD = RAD \left(\text{COSEC } \tfrac{1}{2} ABC - 1 \right)$$

Then mark out a semicircle between them by twisting the chain round from the centre of a line set out at right angles to the tangent lines, which line will join the beginning with the end of the curve. (2) In the case of a radius over one chain in length, join any two points K and L in the tangent lines B A and B C, and calculate the lengths B K and B L in fig. 1 from the following formulæ :—

$$B K = K L \frac{\text{sine B L K}}{\text{sine A B C}}$$
$$B L = K L \frac{\text{sine B K L}}{\text{sine A B C}}$$

Then proceed as in Case 3 (pages 210, 211). If the tangent points F and G be joined by a chord line F G, the curve might also be set out by offsets measured from the chord, as shown in Case 4, fig. 2 (pages 214, 215). The middle point J is found by measuring the length of the chord H G and bisecting it. The distance J M or J K to the points M or K at which offsets are required is next measured, and the square of the distance is subtracted from the square of the radius. The square root of the result is then determined, and from it is subtracted the square root of the radius squared minus half the chord squared. This is expressed in the formulæ given upon the diagram.

If the point of intersection of the tangent lines is inaccessible, the application of the formulæ for finding the distances B K and K L (as above) becomes necessary when Case 3 or Case 4 is to be adopted. Where obstructions occur upon the line of the curve itself, the application of the method shown in Case 4, fig. 2, is often advisable. (Pages 214, 215.)

In fig. 1 (Case 4) the true measurements for setting out a curve are given, the distances along the tangent line from the beginning or end of the curve being respectively equal to the half chords, or the sines of the angles at the centre of the circle for a radius equal in length to a single unit; while the length of the offsets at right angles to the tangent lines are found by subtracting from a unit of measurement the value of the cosines of the angles at the centre of the same circle. If, as shown in fig. 1, the angles taken at the centre of the

SETTING OUT CURVES.

circle be equal,—that is, if the angles taken be subtended by equal chords,—the points marked upon the curve will be at equal distances apart. If $T = $ the distance in links measured along the tangent line by the chain, the offset to be measured by a tape or rod will amount to $R - \sqrt{R^2 - T^2}$ expressed in links. Tables are published in a form suitable for carrying in the pocket which give the length of various offsets at stated distances for different radii.

When taking cross sections upon a curve line already staked out, each cross section has to be set out at right angles to a tangent line. Thus, if in Case 5 (pages 218, 219) a cross section be required at the peg C, the surveyor would measure the length of the chord from the peg at which the cross section is needed to some point, say D, further on in the curve, and he would then calculate and set out the offset required upon the outside of the curve in order to obtain the direction of a tangent line to the point C. Then, by setting up a pole near to the point D at the calculated distance, a right angle to this tangent at C will give a radial line for the direction of the transverse section.

Case 5 illustrates the best method for setting out curves of very large radii with one theodolite by angles at the circumference, and is based upon the fact proved in Case 1, that at any portion of a circular curve the angle subtended by an arc of a fixed length is constantly the same, no matter in what part of the curve the arc or distance be taken, and the angle subtended by any other distance is proportional to that distance. To apply this principle in practice, the late Professor Rankine suggested the measurement of a chord line, which should be taken in comparative short lengths, so as to be practically equal to the length of the curved line adjacent to the length of chord. An accurate circle might thus be completely set out the same as a polygon of an infinite number of sides; but in practice, so long as the length of the curve line is not more than one link greater than the length of chord, the method is accurate. Referring to fig. 1 (pages 218, 219), it will be seen that the length of the chord A B is practically equal to the length of the curve A B, but the length of the chord A C is less than the length of the curve A B C, and the length of

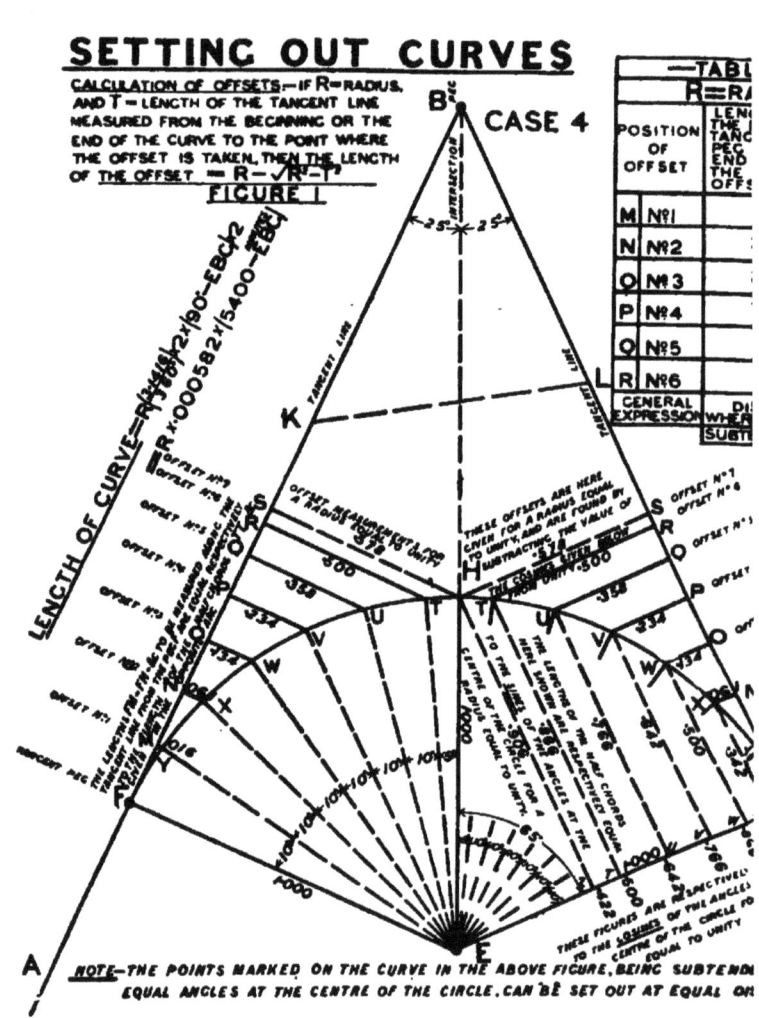

Land Surveying and Levelling, pp. 214, 215.

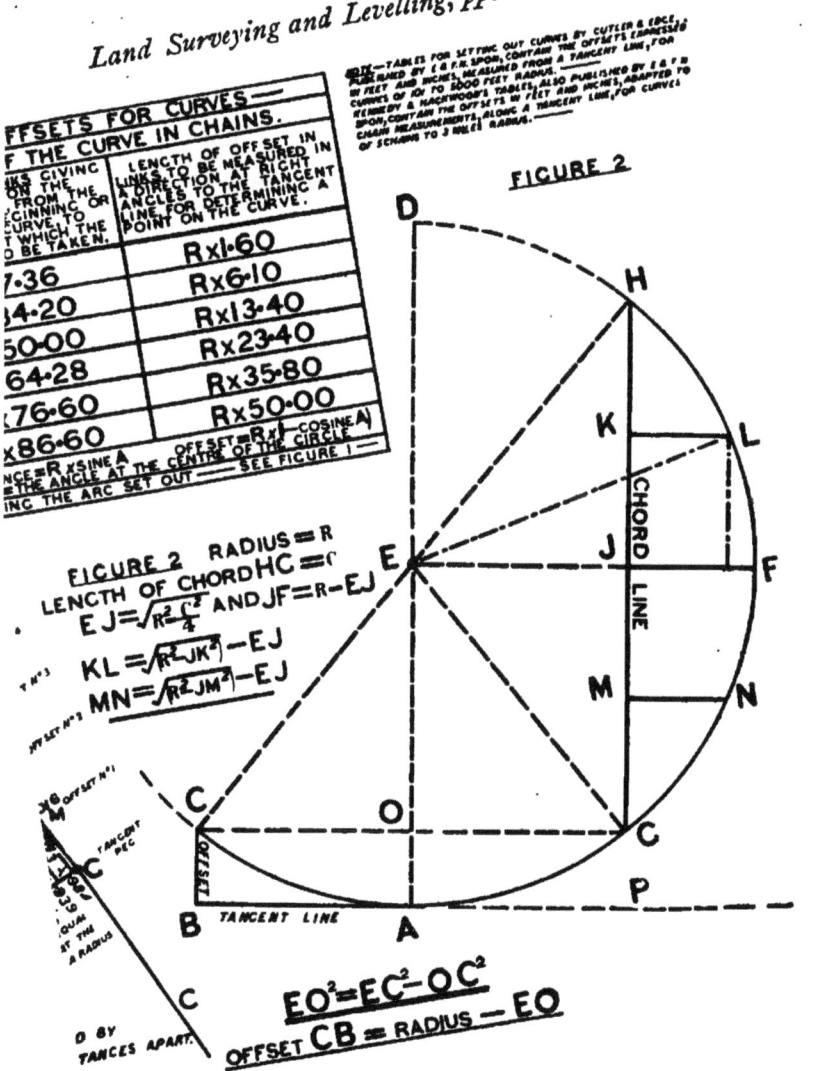

the chord A E is very much less than the length of the curve A B C D E.

The length of a chord line which subtends an angle of 60 degrees at the centre of a circle is equal to the radius; but the length of an arc or curve equal to the radius subtends an angle at the centre of the circle of 57·29 degrees. For a radius of two miles (160 chains) it will be found that the length of a chord of two chains is approximately equal to the length of the curve it connects, but that for a radius of less than one-tenth of this distance (16 chains) it is advisable to take half-chain chords. For curves of less than 5 chains (330 ft.) radius, the arc subtending an angle of 60 degrees at the centre of the circle is so soon approached, and the length of chords to be measured would be so small, that the method shown in Case 3 will furnish the best curve. On the other hand, when Case 3 is applied to curves of over 5 chains radius, the length of the tangent lines from their point of intersection to the commencement and end of the curve is so long, while the perpendicular offsets become so short, that the method explained by Case 5, with the use of a theodolite for setting out the tangential angles, is more advisable.

Hence for a curve off a chord,

Offset $B = \dfrac{\text{chord}}{2R}$ approximately, on diagram (pages 206, 207).

A very convenient list of tangential angles and of angle-multiples for curves, from 5 to 200 units as a radius, has been compiled by Mr. Alexander Beazeley, M.Inst.C.E., and is published as a set, printed on card, for use in the field. The individual card, giving the required tangential angles for any required radius, can be placed on the theodolite in one of the clip spring pieces which are intended to hold the magnifying glass for the reading of the vernier. The angles marked upon the card are calculated from the formula given in Case 1, namely, "Angle in minutes equals the chord divided by radius when the result is multiplied by 1719." This angle is set out with the use of the opposite vernier, as indicated by the small sketch of the plan of a theodolite at the top of our diagram. (See pages 218, 219.)

SETTING OUT CURVES.

We give a copy of the figures upon one of these cards as an illustration. (See Case 5, pages 218, 219):—

Radius 20.

	°	′	°	′
1	1	26	358	34
2	2	52	357	8
3	4	17¾	355	42¼
4	5	43¾	354	16¼
5	7	9¾	352	50¼
6	8	35¾	351	24¼
7	10	1¾	349	58¼
8	11	27¾	348	32¼
9	12	53½	347	6¾
10	14	19¼	345	40¾

Multiples for 20 Rad.

1	85·94367
2	171·38734
3	257·83101
4	343·77468
5	429·71835
6	515·66202
7	601·60569
8	687·54936
9	773·49302

With the use of a book containing the angles required, the angle is read, and the book generally has to be placed away in the pocket, so as to leave both hands free for setting the instrument; but with the use of a special card temporarily fixed upon the theodolite, the angles to be set out are constantly under the Surveyor's inspection.

Messrs. Cutler & Edge's tables are intelligible even to a workman who may not understand the formula upon which they are based. The book is clearly printed and of a convenient size to carry about as a pocket book.

When the theodolite is not divided to the same fractional parts of a degree given in the printed list of tangential angles, these angles may be set out to the nearest subdivision upon the vernier to the amount furnished by the tables. If, as is usual in railway work, the radius of the curve is expressed in chains of Gunter's links, the same chain must be employed for the measurement of the chord, otherwise any other unit measurement can be employed

SETTING OUT CURVES
— WITH THE USE OF A THEODOLITE —

NOTE - THE THEODOLITE IS SET UP OVER THE POINT MARKED A AT THE JUNCTION OF THE CURVE WITH THE STRAIGHT LINE, AND THE VERNIER UPON THE HORI-ZONTAL PLATE CLAMPED TO 360° OR ZERO. — THE CROSS HAIRS IN THE TELESCOPE ARE THEN DIRECTED TOWARDS ANY PEG UPON THE TANGENT LINE, THE LOWER LIMB OF THE INSTRUMENT IS CLAMPED WITH THE CROSS HAIRS ACCURATELY SET ON THE TANGENT LINE, THE VERNIER PLATE UNCLAMPED, AND THE TANGENTIAL ANGLES SET OUT AS REQUIRED. —

CASE

-RANK

TANGENTIAL ANGLES FOR 1 CHAIN CHORDS

RADIUS OF CURVE CHAINS	TANGENTIAL ANGLE DEG MIN	RADIUS OF CURVE CHAINS	TANGENTIAL ANGLE DEG MIN	RADIUS OF CURVE CHAINS	TANGENTIAL ANGLE DEG MIN
5	5 - 43·8	20	1 - 25·9	50	- 34·4
8	3 - 34·8	25	1 - 8·7	60	- 28·7
9	3 - 11·0	30	- 57·3	70	- 24·6
10	2 - 51·9	35	- 49·1	80	- 21·5
12	2 - 23·2	40	- 42·9	160	- 10·7
15	1 - 54·6	45	- 38·2	240	- 7·2

IN THE ANNEXED TABLE, THE UNIT OF LENGTH FOR A RADIUS, IS OF THE SAME DENOMINATION AS THE UNIT OF LENGTH EMPLOYED FOR THE CHORD: SO THAT THE SAME CHAIN, MUST BE USED IN ANY SINGLE CURVE, FOR THE MEASUREMENT OF THE CHORD, AS IS ASSUMED FOR THE RADIUS. — IN RAILWAY WORK, GUNTER'S CHAIN IS MOST GENERALLY EMPLOYED: — CURVES OF OVER TWO MILES (160 CHAINS) RADIUS, CAN BE SET OUT ACCURATELY BY TWO CHAIN-CHORDS, & ACCOMPLISHED BY MEASURING THE CHORDS WITH A COUPLE OF CHAINS TIED TOGETHER, AND TAKING AS THE UNIT ANGLE DOUBLE THE VALUE OF THE TANGENTIAL ANGLE FOR A SINGLE CHAIN CHORD.

CURVES OF LESS THAN 15 CHAINS RADIUS SHOULD BE SET OUT IN HALF CHAIN-CHORDS — THE UNIT ANGLE FOR HALF CHAIN-CHORDS IS HALF THE ANGLE FOR 1 CHAIN-CHORDS.

NOTE - BEASLEY'S TABLES OF CURVES, A SET PRINTED ON CARDS, ARE MOST CONVENIENT FOR USE IN THE FIELD, AS THE SPECIAL CARD GIVING THE TANGENTIAL ANGLES FOR THE REQUIRED RADIUS, CAN BE PLACED ON THE THEODOLITE, THUS LEAVING BOTH HANDS FREE FOR SETTING THE INSTRUMENTS FOR SIGNALLING. THEY ARE PUBLISHED BY CROSBY LOCKWOOD & CO., LONDON. SIMMS & MACKWOOD'S TABLES, A SMALL BOOK PUBLISHED BY E & F.N. SPON, LONDON ALSO CONTAINS THE TANGENTIAL ANGLES REQUIRED FOR SETTING OUT CURVES OF FIVE CHAINS TO THREE MILES RADIUS WITH CERTAINTY AND EXPEDITION. —

BASE LINE OR CENTRE LINE OF WORK . A TANGENT PEG

THE TANGENTIAL ANGLE FOR A CHORD CONSISTING OF ANY NUMBER OF LINKS, IS THE TANGENTIAL ANGLE GIVEN FOR A CHORD ONE CHAIN IN LENGTH MULTIPLYING THE TANGENTIAL ANGLE GIVEN FOR A CHORD ONE CHAIN IN LENGTH HUNDRED LINKS, CONTAINED IN THE LENGTH OF CHORD REQUIRED.

Land Surveying and Levelling pp. 218, 219.

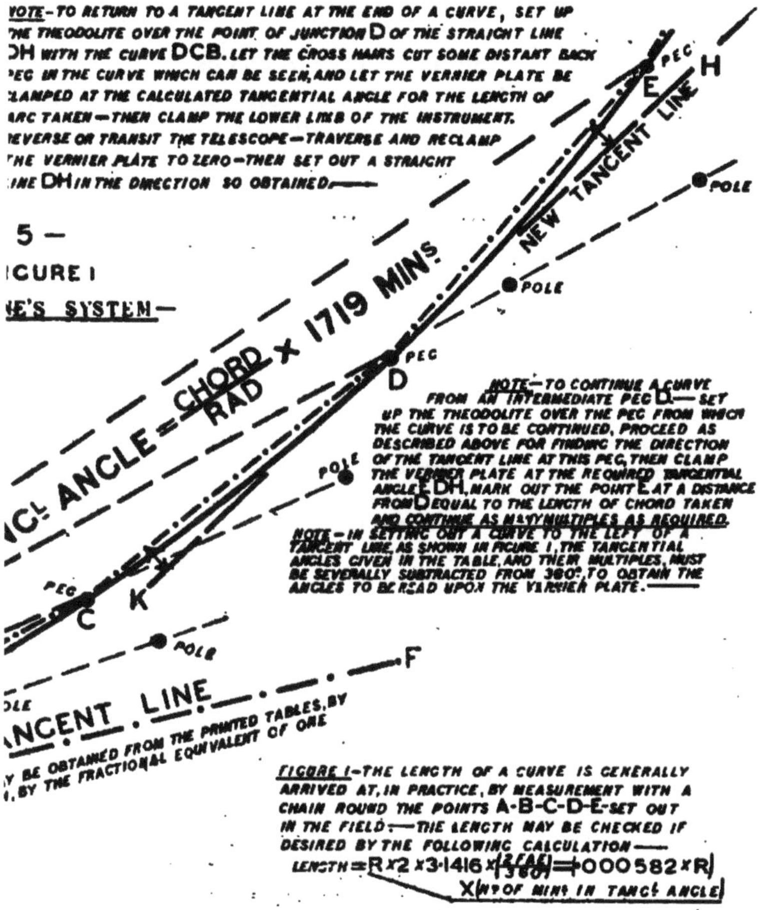

NOTE–TO RETURN TO A TANGENT LINE AT THE END OF A CURVE, SET UP THE THEODOLITE OVER THE POINT OF JUNCTION D OF THE STRAIGHT LINE DH WITH THE CURVE DCB. LET THE CROSS HAIRS CUT SOME DISTANT BACK PEG IN THE CURVE WHICH CAN BE SEEN, AND LET THE VERNIER PLATE BE CLAMPED AT THE CALCULATED TANGENTIAL ANGLE FOR THE LENGTH OF ARC TAKEN—THEN CLAMP THE LOWER LIMB OF THE INSTRUMENT. REVERSE OR TRANSIT THE TELESCOPE—TRAVERSE AND RECLAMP THE VERNIER PLATE TO ZERO—THEN SET OUT A STRAIGHT LINE DH IN THE DIRECTION SO OBTAINED.
5 —
CURE I
E'S SYSTEM—
CL ANGLE = CHORD / RAD x 1719 MINS
NEW TANGENT LINE
PEG
E
H
POLE
POLE
PEG
D
POLE
PEG
C
K
POLE
F
NCENT . LINE
Y BE OBTAINED FROM THE PRINTED TABLES, BY
, BY THE FRACTIONAL EQUIVALENT OF ONE
NOTE—TO CONTINUE A CURVE FROM AN INTERMEDIATE PEG D.—SET UP THE THEODOLITE OVER THE PEG FROM WHICH THE CURVE IS TO BE CONTINUED, PROCEED AS DESCRIBED ABOVE FOR FINDING THE DIRECTION OF THE TANGENT LINE AT THIS PEG, THEN CLAMP THE VERNIER PLATE AT THE REQUIRED TANGENTIAL ANGLE EDH, MARK OUT THE POINT E AT A DISTANCE FROM D EQUAL TO THE LENGTH OF CHORD TAKEN AND CONTINUE AS MANY MULTIPLES AS REQUIRED.
NOTE—IN SETTING OUT A CURVE TO THE LEFT OF A TANGENT LINE, AS SHOWN IN FIGURE I, THE TANGENTIAL ANGLES GIVEN IN THE TABLE, AND THEIR MULTIPLES, MUST BE SEVERALLY SUBTRACTED FROM 360°, TO OBTAIN THE ANGLES TO BE READ UPON THE VERNIER PLATE.
FIGURE I–THE LENGTH OF A CURVE IS GENERALLY ARRIVED AT, IN PRACTICE, BY MEASUREMENT WITH A CHAIN ROUND THE POINTS A-B-C-D-E-SET OUT IN THE FIELD:—THE LENGTH MAY BE CHECKED IF DESIRED BY THE FOLLOWING CALCULATION—
LENGTH = R x 2 x 3·1416 x (2(AE)/360) = ·000582 x R
X (NO OF MINS IN TANGL ANGLE)

with the use of these tables, provided the same denomination is taken for the chord as is assumed for the radius. The tangential angle in minutes for 100 ft. chords when the radius is expressed in chains of 66 ft.

$$A° = \frac{2604\cdot 4}{\text{Radius of curve in chains}}.$$

The tables of multiples give the tangential angle in minutes and decimals for units of radius up to 9, and are intended to facilitate the determination of the tangential angle for fractional chords (page 217). Thus if a curve of 20 chains radius commence at ... miles, ... furlongs, and 37 links from the starting point, the tangential angle for the fraction 63 links will be ascertained thus :

·60 = 51·566202
·03 = 2·5783101

Angle for 63 links = 54·1445121 or $\frac{63}{100}$ of 1° 26′ = 54 minutes.

The length of chord of 63 links is then set out at an angle of 54 minutes with the tangential line, after which the unit angle for chain chords is added to this value at each setting out of points one chain apart. Again, if a curve is to terminate at 63 links beyond a full chain measurement, this value is to be added to the tangential angle taken for the last whole chord of the curve. To return to a straight line, as at D in fig. 1 (pages 218, 219), the theodolite is set up over this point, and any previous point along the curve C, B, or A is selected, the distance of which measured by chords is known. The instrument is clamped to the tangential angle for this distance, and the telescope is directed to this point and the lower plate clamped. The vernier is then unclamped and set back to 360°, when the telescope will be found to be in the direction of the tangent line D K, and when traversed vertically to be in the direction of D H. If, as in fig. 1, the vernier for a radius of 20 units has been successively set to 358° 34′ – 357° 8′, 355° 42′ for pegging out respectively the points B, C, D, when the tangent line is to the right hand of the curve, we must remember that the tangent line D K being to the left of the curve when the instrument is set up at D, the point A must be viewed with the vernier clamped to 4° 17′ or the point B with the vernier clamped to 2° 52′, and then the direction of a line joining the 360° or zero point with the centre of the instrument will give the direction D K. (Pages 218, 219.)

The poles shown in the direction of the chord lines A B, A C, A D (Case 5) are not generally necessary. The usual method is for one man to hold one end of the chain at the last point determined, taking care, if the curve be flat, to place his body upon the outside of the curve, so as not to impede the line of sight when the theodolite is set for fixing the next point in the curve. The other assistant pulls out the chain or the tape to the given length, and holds up a peg or lath, which he keeps vertical at the correct distance, moving it about as directed by the surveyor, to the right or left hand, until it accurately appears in the required direction. Should any obstacle render it necessary to remove and reset up the theodolite over a new point in the curve, the direction of a new tangent line must be found by the method shown in fig. 1 (Case 5), and the same process of setting out by means of tangential angles re-commenced. The use of the tangential angles, which are calculated from the formula proved by Case 1, enables the curve to be set out to the right-hand side of the tangent line, when the theodolite is placed over the beginning of the curve, as the primary scale of divisions upon the horizontal circle of the instrument is numbered to read in the direction of the hands of a watch; hence when the curve is to be set out to the left-hand side of a tangent line, the column upon the card containing the differences of the tangential angles must be adopted. Thus with a radius of 20 chains, if 1° 25′ 57″ be the tangential angle for a chord of one chain in length, and an angle of 2° 51′ 54″ be the tangential angle for the intersection of a second chord of one chain's length round the arc, when the curve is to be set out to the right-hand side of the tangent line; then the tangential angles to be employed for setting out two points at the same distances for a curve to the left-hand side of the tangent line will be 358° 34′ 3″ and 357° 8′ 6″ respectively.

Fig. 2 (pages 222, 223) illustrates a method of setting out

CIRCULAR CURVES
RELATIVE VALUE OF USEFUL ANGLES TO ANGLE OF INTERS[ECTION]

FIG 2

LET X = HALF THE ANGLE OF INTERSECTION
THEN $A = 90° - X$
$B = 90° - X$
$C = 90° + X$

$D = 90°$
$E + F = 90°$
$G + H = 90°$
$R = T \times T$

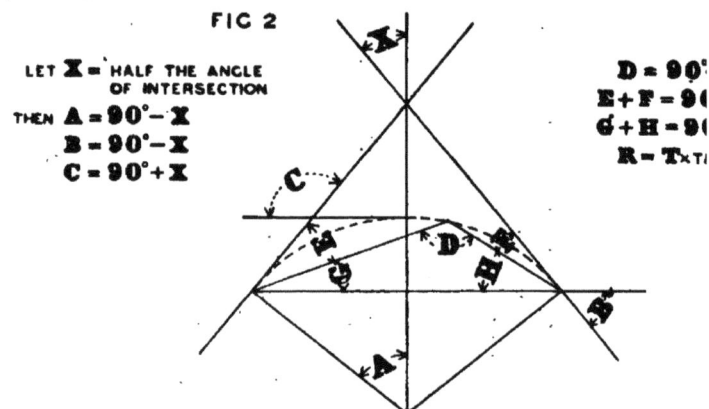

SETTING OUT WITH TWO THEODOLITES

IF ANY TWO LINES BE SET OUT FROM THE STARTING POINTS OF THE CURVE HAVIN[G] THEIR TANGENTIAL ANGLES $(E + F) = 90° - X$, THE INTERSECTION OF THESE TWO LIN[ES] POINT IN THE CURVE — BY A SERIES OF SUCH POINTS THE WHOLE CURVE MAY BE SE[T] COUNTRY BE SUFFICIENTLY OPEN TO ALLOW OF IT — WHEN THE COUNTRY IS UNEV[EN] SHOWN IN FIG 1 BY SHIFTING AND RE-SETTING UP THE THEODOLITE WHERE NE[CESSARY] MORE APPLICABLE —

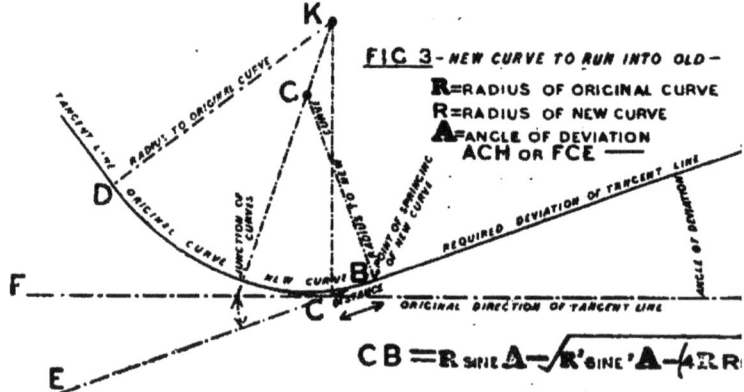

FIG 3 - NEW CURVE TO RUN INTO OLD —

R = RADIUS OF ORIGINAL CURVE
R' = RADIUS OF NEW CURVE
A = ANGLE OF DEVIATION
ACH or FCE ——

$CB = R \sin A - \sqrt{R'^2 \sin^2 A - (4RR')}$

Land Surveying and Levelling, pp. 222, 225.

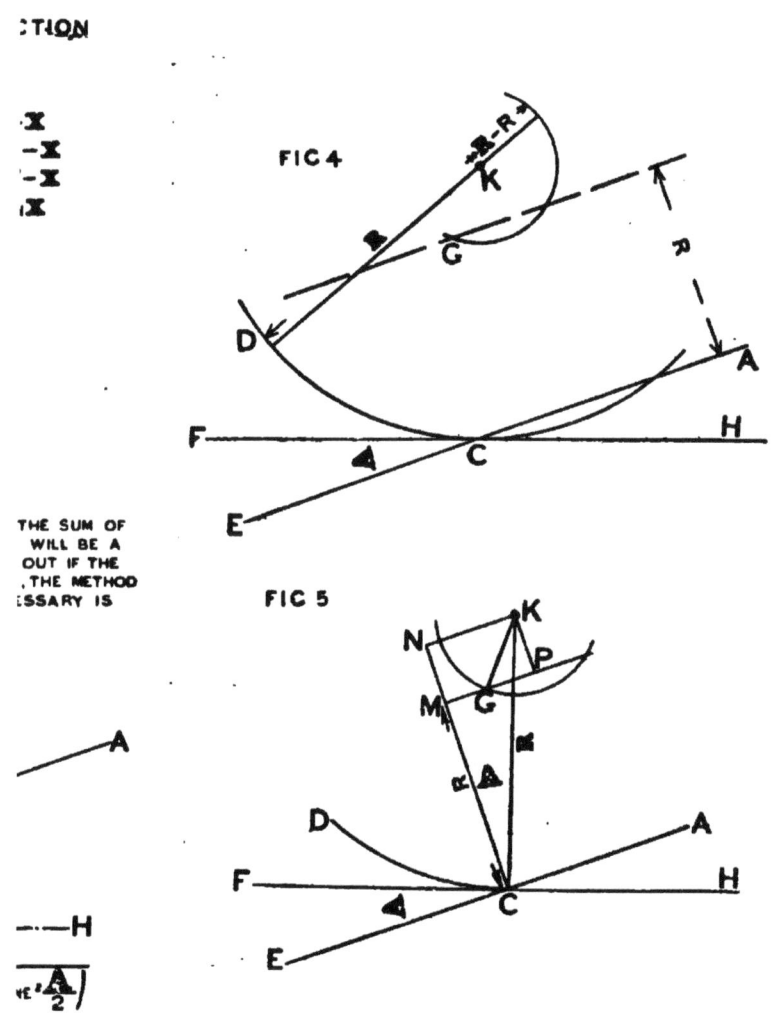

curves with the use of two theodolites, each set up respectively over points at the beginning and the end of a curve, when the angle of intersection between the tangent lines which pass through those points is known.

In-road work, curves may be set out by offsets from lines drawn upon a plan, and afterwards be checked by means of offsets from the tangent lines. When in road-work it is desirable to accurately set out one curve running into another, the formula given in fig. 3 may be employed. The proof is given in figs. 4 and 5 (pages 222, 223).

$$C B = R \sin. A - \sqrt{R^2 \sin.^2 A - 4 \left(R r \sin.^2 \frac{A}{2} \right)}$$

If $R = 7$ and $r = 5$ when $A = 20°$
$\sin. A = \sin. 20° = ·342$
$R \sin. A = 7 \times ·342 = 2·394$

$\sin.^2 A = ·116964$
$R^2 \sin.^2 A = 7 \times 7 (·116964)$
$\qquad = 5·731376$
And $\sin. \frac{A}{2} = \sin. 10° = ·173$

$\sin.^2 \frac{A}{2} = ·029929$

$4 R r (·029929) = 4·190060.$

The square root of the difference between $5·731376$ and $4·190060$ equals $1·241$.

And $(2·394 - 1.241) = 1·153 = C B$, the distance required. The geometrical proof is shown in fig. 4. The algebraical proof is illustrated by fig. 5. (See pages 222, 223).

Fig. 4.—In order that the circle with the centre G, in fig. 3, should touch the tangent line E A, its centre must be in a line parallel to it, at a distance equal to its radius (see fig. 4), and in order that it should touch the circle having its centre at K, its centre must be in a circle having a radius equal to the difference of the two given radii, concentrical with the circle D C, which touches the tangent line F H (fig. 4). Therefore it lies in the intersection of the parallel line and the circles with a radius equal to the

SETTING OUT CURVES.

differences of the two given radii. This intersection comes in the point G, fig. 4 (pages 222, 223). The problem, therefore, algebraically resolves itself into the following:—

If A = the angle $F C E = A C H = N C K$ (fig. 5),
  R = radius of orignal curve,
  r = radius of new curve,
then $N C = R \cos. A$,
  $M C = r$,
  $M G = N K - G P$,
  $N K = R \sin. A$,
  $G P = \sqrt{KG^2 - KP^2}$
  $K G^2 = (R - r)^2$
  $K P^2 = (N C - M C)^2 = (R \cos. A - r)^2$
$\therefore G P = \sqrt{(R - r)^2 - (R \cos. A - r)^2}$
$ = \sqrt{R^2 - 2Rr + r^2 - R^2 \cos.^2 A + 2Rr \cos. A r^2}$
$ = \sqrt{R^2 - (1 - \cos.^2 A) + 2 Rr (\cos. A - 1)}$
$ = \sqrt{R^2 \sin.^2 A + 2Rr(1 - 2\sin.^2 \frac{A}{2} - 1)}$
$ = \sqrt{R^2 \sin.^2 A - 4Rr \sin.^2 \frac{A}{2}}$

$\therefore M G = R \sin. A - \sqrt{R^2 \sin.^2 A - \left(4 Rr \sin.^2 \frac{A}{2}\right)}$

CHAPTER XIX.

MARINE SURVEYING.

In Marine Surveying, the use of two theodolites upon the shore can be employed for recording the position of a sounding by the intersection of lines making known angles with a given base-line. When the box sextant is used, the angles are taken at the points marked A, B, and C by the surveyor in a boat, these points being shown in the above diagram,

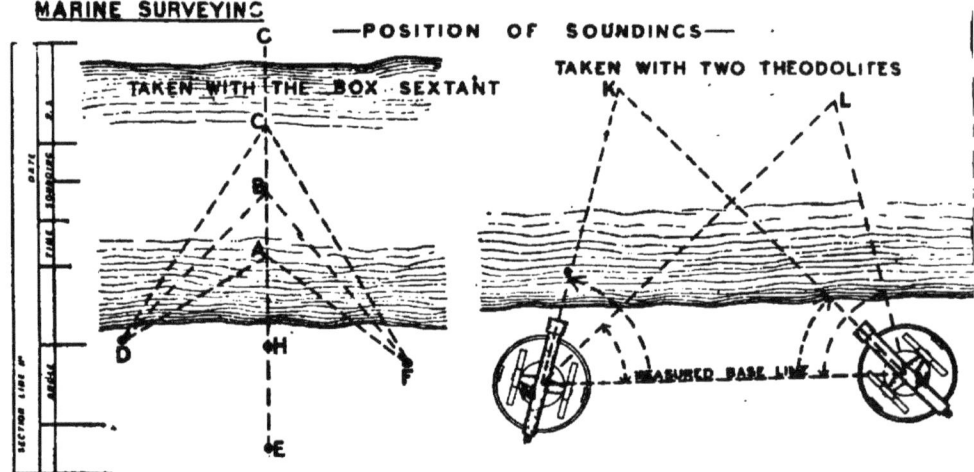

and the angles bear upon the stations D and E, and between the stations F and E, as near upon the line of section G E as wind and tide will permit.

A good form of book for recording soundings consists of pages of ruled line paper with the addition of vertical lines dividing a page into five columns so as to enter when taking a section along a line, say between H and G as shown in the figure. (Page 226.) The records give: (1) distance from H obtained by tallies fixed to a stranded copper wire attached to the shore at H, and attached round a reel in the centre of the boat, the wire being pulled taut by the man who winds the reel by means of a crab handle and by the oarsman rowing the boat so as to render the wire in tension; (2) a column to enter the time, which it will be sufficient to enter every five minutes; (3) the record of a gauge to show height of water, an operation meanwhile performed by another man at some fixed station near at hand, who enters the readings in an independent book, so as to be subsequently repeated in the sounding book; (4) the sounding read in feet and inches off a sounding chain usually 80 feet long, in which the links measure an inch; (5) the reduced sounding for plotting. This operation requires a man at H to sight a mark at G, and to direct the boat in line or preferably to fix a back station at E, so that the men in the boat can place themselves in line with H and E, and in the boat besides the surveyor, two men at the oars, one man at the sounding, and one man to wind the reel or let out the stranded wire as directed. The columns of a book suitable for a section are further explained in the form given on page 229. The soundings upon the Admiralty Charts are expressed in feet for the mean low-water of spring tides, and reference should be made to the diagram of the compass card on these maps, showing the magnetic variation of the direction of north from the true north. The surveyor should also make himself familiar with the readings of the compass-card, so as to feel quite at home with the boatmen whose language involves the continual use of such descriptions of wind and direction.

By means of a current meter, the rate of flow in any river or stream, or in a tidal current, can be readily and accurately ascertained, either at or below the surface, as may be required. The meter shown on page 228 is a considerable improvement upon instruments of this kind hitherto in use, the parts most exposed to wear being thoroughly strengthened, and the indication of flowing being at once shown upon the dial. It consists of an axis carrying two screw blades, and having an endless screw, which works a differential disc in two parts, the front part showing the number of feet run, from 0 to 330; and the back part, miles and furlongs, from 0 to 12 miles. The arm carrying this disc is so attached to the frame that the disc, when not in use, is kept free of the endless screw by a coiled spring, but can be instantly brought into working contact with it, either by a cord held by the operator, or by a nut on the vertical screw at one end of the arm. A flat vane, clamped to the frame, serves to keep the instrument with its screw properly facing the current.

The mode of using the current meter is as follows:—

The meter is to be well fitted, and securely clamped to the end of a stout stick—preferably a cane—the flat

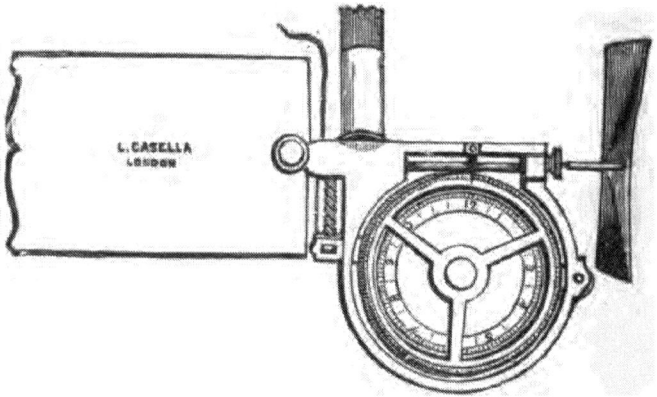

vane to be fixed in place, and a cord attached to the eyehole of the vertical screw. The differential disc is then to be set to zero of each graduated circle, in which position it will be held by a point projecting on the inside of the bottom of the frame. The meter is then to be lowered into the water, the screw or turbine being held facing the current, and when the intended depth is reached, the differential disc is to be brought into contact with the endless screw or shaft by tightening the cord, the time being carefully noted at the moment of contact, and again on releasing the disc. The distance registered on the graduated circles,

MARINE SURVEYING.

being divided by the time, gives the rate of flow per minute or per second, as required.

If, however, time can be given for a long run, and especially in a rapid current, the disc may be brought into contact by screwing up the small nut under the eye-hole before immersion, the time being taken at the moments of entering and leaving the water.

In a rapid current the vibration caused by the screw is considerable; precaution must, therefore, be taken to hold the instrument firmly in its true position, and when the observation is taken at some depth it will be found useful to have a tie head at the top of the staff for this purpose.

It is also advisable to attach a line to the eye-hole of the vertical screw, whether contact be made before entering, or in the water, in case of the instrument becoming by any accident detached from the stick, mentioned above.

The Station Pointer (page 81) is used for plotting the position of any soundings A, B, C, shown in the figure, and such position is fixed by simultaneously taking the bearings of three known objects on shore D, E, F, or in the case of a river section upon the line E G, the bearing of two objects D and F right and left of that line. The position of the point may be found by means of tracing paper, on which observed angles are plotted, in which case the draughtsman shifts the tracing paper, with the angle plotted thereon, over the plan until the point falls on to the only spot from which the lines containing the observed angles will intersect the three fixed stations. (See illustration, page 226.)

Section Line No............ Date.....................

Column for any general remarks.	Angle.		Time.	Sounding.	R. S.
	In this column, the angles DCF, DBF, DAF, are to be entered.	In this column, the description of the points D and F on shore is given.			In this column, the reduced soundings are entered, being the heights above the given datum.

K

The form of book usually adopted is, as shown, a record of outdoor entries, headed with a description of the line upon which the section is taken and the date. The first column upon each page is intended for any general remarks, the next column to give the angles D C E and E C F, D B E and E B F, D A E and E A F. In the third column the description of the points upon the shore D, E, and F is stated. Then the time, and the soundings are entered under their heading, while the last column entitled "R. S." contains the entries of the reduced soundings, being the heights above a given datum.

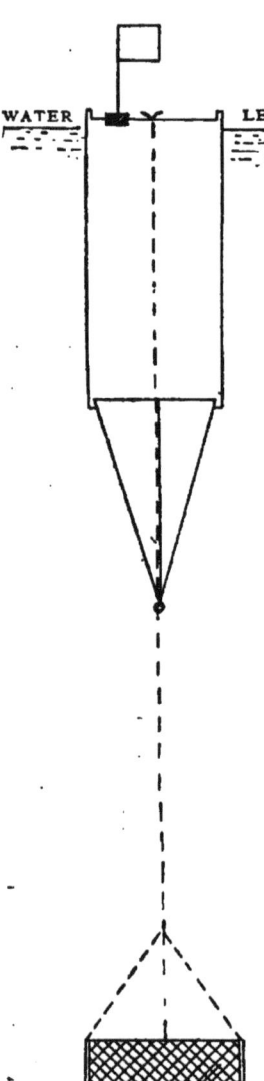

In the case of a tidal current survey the floats used may consist of a thin galvanized sheet iron cylinder about 12 in. to 15 in. diameter, closed at both ends and thoroughly watertight, provided at the top with a rope twist and a bung hole stopped with a cork carrying a small flag, while underneath are attached 3 legs made of ⅜-in. iron meeting in a ring. (See sketch.) A strong cord marked in fathoms passes from the rope twist through the ring to a wicker basket carrying about 50 lbs. of ballast. The cord connected between the bottom ring and the rope twist passes outside the cylinder, which tends to twist it slightly out of the vertical, but this is unimportant as the plotting scale is so small a ratio,

The floats should not be set deeper than 6 ft. from the bottom. It is well to start at 9 ft. from the bed of the river and to adopt a clearance of not less than 6 feet or more than 12 ft. The depth in the course of the float can be tested by the sounding line in the boat, and the depth of the basket altered accordingly so as to avoid touching the bottom. If the basket touches in a strong current, the float will go under and perhaps be lost. To prevent risk of this, the surveyor should coil the spare end of rope neatly on the top of the drum and fasten a cork to the extreme end of it; this may float and so allow the drum to be recovered.

To obtain the surface velocity, a slice of turnip may be thrown in close to the float and its velocity relative to the float noted by the surveyor. Each float is attended by a small rowing boat carrying two boatmen and a surveyor, one man manages the boat and keeps it near the float, the other attends to the soundings and takes the depth of the basket under his care, altering it as the soundings require. The surveyor takes all angles for fixing the position of the float and books all observations of sounding, etc. The field book may be kept somewhat as follows:—

Time	Object.	Angle.	Object.	Sounding.	Remarks.
In hours and minutes.	On which the angle is taken (direct).	In degrees and minutes.	On which the angle is taken (reflected)	In fathoms and quarter fathoms.	As to check observations, crossing lines, proximity of buoys, surface velocity, basket fouling, etc., etc.

Observations of the position of a float can be taken with advantage every fifteen minutes, commencing at the hour, and the exact time of passing four or five well defined lines running across the course taken by the float should be recorded. The depth at which the basket hangs should be noted, and the observations continued till the turn of the tide, or until some pre-arranged time or point is attained. A sketch of the principal landmarks and of each of the buoys should be made in the field book, as without this, the record of all observations would be vague.

The apparatus required on each boat would be as follows:—(1) A nautical sextant; (2) A watch, preferably one showing seconds; (3) A pair of field-glasses; (4) A chart of the water to be traversed; (5) A pocket compass for safety in case of fog; (6) A sounding lead line; and (7) A few turnips.

The lead, or sounding line, employed for nautical purposes consists of cord made of material combining the qualities of toughness with flexibility, and is usually marked to furnish fathoms and to read to quarter fathoms. Some lines are, however, marked to lengths of five feet and read to feet and tenths of a foot, with the aid of a foot rule or staff, which are occasionally of service for engineering surveys. The sounding line is loaded with a conical lead weight which may be armed with tallow in the hollow of its base, when it is required to bring up a specimen of the soil at the site; but on hard material, a jagged plunger should be employed. The lead line or sounding cord needs to be periodically tested with a standard measure during the progress of a long survey, whenever accuracy of depth needs to be very carefully recorded; it is advisable to test it daily before use. The nature of the bottom should be frequently noted, especially marking any sudden changes that may occur in the depth, as such variations in soundings usually indicate the proximity of neighbouring shoals. When taking soundings the surveyor should always try to steer in a straight line, the ends of which line should be fixed in position for indication on the plan, and the surveyor should endeavour to keep two well-defined objects in view as one point upon a continuation of this line, one near and one as far off as possible. The writer says "endeavour" to do so, because it will be found impossible with a tidal stream and wind to keep accurately upon the line, and the angles taken by the sextant will then indicate the exact position of sounding right or left of such line, but the nearer thereto the better for the plan. Soundings should not be taken during very high tides, and should be made during neap in preference to spring-tides, and in ebb in preference to flood-tides. Soundings to be taken in flood-tides, especially during springs, should not be made till within an hour of high-water, but the ebb-tide will, as a rule, be found more steady than the flood-tide. The high-water can be timed from a mean of the time of continuance of the highest reading upon the tide gauge.

The starting position for these investigations is primarily fixed in the same manner as all other positions in a nautical survey, that is by two angles on three known fixed points, but being once found, two lines inclined as nearly as possible at right angles are obtained; the approximate position of the boat can thus be discovered at any future day without any observation with the sextant.

In a tidal navigation survey that was upon one occasion being made at the mouth of the Thames it was found that a current running at $1\frac{1}{2}$ to 2 miles per hour was not sufficient to sink the float when the basket grounded. The spare rope, with the cork at its extremity, will not remain on the top of the float unless the sea is very smooth.

The small flag, shown in the diagram (page 230), on the float is not needed on all occasions, and is liable to "bob" into the boatman's face when altering the depth of the basket.

The sextant remains in a much better condition if it is kept in the box when not actually in use, and it is frequently advisable to use the "dummy" in preference to the telescope, as such a small field is covered by the latter.

The surveyor should always take the more indistinct of two angles first, and look "directly" at the most indistinct object—if this is on the observer's right-hand side he holds the sextant upside down. Also when taking angles, he should have the boat's nose kept straight on to the waves, as less rocking takes place in that position. In fogs, and when no better objects are visible, ships at anchor may be made use of, and their position fixed later; but, as a rule, distant points are better than points close at hand, because the angles do not change so rapidly. At the same time, they are not so easily seen under varying conditions of atmosphere.

CHAPTER XX.

PLANS FOR DEPOSIT.

In the instructions given by the Commissioners of the Copyhold Inclosure and Tithe Office for the preparation of first-class plans, it was provided that the Commissioners did not pledge themselves to seal a plan to which any of the following objections applied without testing it upon the ground :—(1) Where there was any reason to distrust the authenticity or integrity of the Survey; (2) where the means afforded were insufficient to prove the accuracy of the work in all its details; (3) where the plan did not agree with the field-books; (4) where the field-books had been kept in common or metallic pencil; (5) where erasures had been made in the field-books; (6) where alterations had been made in the field-books without a satisfactory explanation being afforded; (7) where the offsets exceeded a chain in length.

The Commissioners appointed the surveyor to be employed when required for testing the plans upon the ground, and required that a sum of money sufficient to cover the expenses should be lodged in their hands before the testing was commenced.

The lines which the Commissioners required to have measured on the ground for testing the accuracy of the plan were defined as three lines in the form of a well-shaped triangle, with a proof line from one of the angles to the opposite side.

The scale (3 chains to 1 inch) employed is explained on page 128.

In measuring the testing lines, all intersections of fences were to be noted, and offsets taken within the ordinary limits of a chain's length.

Distances, measured along the fences joined, were to be

given in writing to all junctions of fences which came within two chains of the test lines upon the plan.

The fields in which the angular points of the triangle occurred were to be wholly surveyed, or enough of their boundaries ascertained to determine precisely the position of those angular points of intersection on the plan.

The entries in the field-books were to be made with ink in the field, and any alterations which required to be made in them were to be attested by the initials of the surveyor. An explanation of the cause of the alteration was also to be entered as a note upon the plan or in the field-book.

A projection of the testing lines on the scale of the original plan was to be sent with the field notes to the office of the Tithe Commissioners.

The cost of testing on the ground did not, as a rule, exceed £15 for a parish or district under 3,000 acres; and for a parish of larger size was reckoned at the rate of about £5 per thousand acres, not including travelling expenses.

Although the above regulations have since the authority given to the Board of Agriculture become obsolete, they contain hints useful for adoption, except the requirement to make the field-book entries in ink, for the reason given upon page 46.

The following particulars should be borne in mind by a surveyor when preparing plans for deposit.

In new roads across fields, the surveyor should aim, in setting out, to have as great a length as possible straight, and as far as practicable cross roads at right angles to one another, as this arrangement not only facilitates the division of land, but contributes greatly to the economy of cultivation with the plough. All old roads that may be deemed unnecessary may be stopped up and allotted to the different claimants or diverted into more convenient directions at the discretion of the surveyor.

In the metropolis all plans for the formation of new streets are required to be made to a scale of 88 feet to the inch, and to be accompanied by longitudinal and cross sections to a scale of 88 feet to the inch horizontal, and 11 feet to the inch vertical, showing the natural and intended surfaces of the streets, and also by a key-plan of the locality.

The levels are to be computed from Ordnance or some other fixed datum, which must be clearly stated.

Forty feet at the least is the width of every new street intended for carriage traffic; twenty feet at the least is the width of every new street intended only for foot traffic; and the said widths, respectively, are construed to mean the width of the carriage and footway only, exclusive of any gardens, forecourts, open areas, or other spaces in front of the houses or buildings erected or intended to be erected in any street.

The measurement of the width of every new street is taken at a right angle to the course thereof, half on either side from the centre or crown of the roadway to the external wall or front of the intended houses or buildings on each side thereof; but where forecourts or other spaces are intended to be left in front of the houses or buildings, then the width of the street, as already defined, is measured from the centre line up to the fence, railing, or boundary dividing or intended to divide such forecourts, gardens, or spaces from the public way.

The carriage-way of every new street should curve or fall from the centre or crown thereof at the rate of three-eighths of an inch, at the least, for every foot of breadth, or at an average transverse gradient of 1 in 32.

In every new street the kerb to each footpath should not be less than four nor more than eight inches above the channel of the roadway, except in the case of crossings, paved or formed, for the use of foot-passengers; and the slope of every footpath towards the kerb must be half an inch to every foot of width if the footpath be unpaved, or not less than a quarter of an inch to every foot of width if the footpath be paved.

Building plots should always be measured in feet and inches, and in every case the frontage and depth should be figured on the plan. Care should be taken that no portion of any new work encroaches upon the remainder of the estate by any overhanging eaves, barges, spouting, stone weathering, etc. (See page 122.)

When called upon to fix a frontage upon the ground, first measure off the length of the piece to be dealt with or pegged out, and then measure the remaining portion of the boundary, as a proof that you have accurately marked off no more nor less than as indicated on your plan.

For the preparation of Street Plans to be submitted to a Local Authority, the following data may be taken to include all that is generally necessary to be indicated.

1.—The heading to state to which class the street belongs.
2.—The plans to be numbered.
3.—The sections to show the ground floor and cellar floor levels of every house in the street and the existing surface of the centre of the street above Ordnance datum.
4.—The cellar and floor levels and also the existing and intended levels of the centre and ends of the street shall be shown on the plan. The levels to the existing surface to be in Black—intended levels in Red.
5.—The levels for the section shall be commenced always at the lowest end of the sewer proposed to be laid in the street.
6.—All reference to floor and cellar levels on right side of the section, *i.e.*, on the right side of the section running from the lowest to the highest end of the sewer, shall be in *Blue*, and on the left side in *Red*.
7.—The existing surface of the street shall be shown by a black line. The intended surface and the bottom of road foundation by red lines.
8.—Existing sewers shall be shown in Black.
9.—Intended sewers in Red.
10.—Surface water sewers in Blue.
11.—The scale for plans made in the Local Surveyor's office to be the same as the Town Plan (41·66 feet = 1 inch). Plans submitted for approval not less than 40 feet to an inch. The vertical scale shall be 8 feet to an inch.
12.—The Horizontal and Vertical Scales for cross sections shall be the same.
13.—The average level of the equinoctial tides above Ordnance datum to be stated where necessary.
14.—Where the level of any street is less than a prescribed number of feet above Ordnance datum, the equi-

noctial tide level shall be shown by a fine blue line drawn across the section.

15.—The scales to be drawn on the plans and figured dimensions to be fully given.

16.—The distance between two manholes or a manhole and a ventilator should not exceed 300 feet. In the case of a change of gradient between two manholes or a manhole and a ventilator, a lamphole shall be built at the point where such change is made. The lamphole shall be 18 inches square inside. The brickwork to be 9 inches thick one foot above the top of the sewer, and $\frac{1}{2}$ brick thick for the remainder of its height. The brickwork to be in Portland cement, and covered with an approved cast-iron cover.

17.—Gullies should not be more than 100 feet apart where the gradient is less than 1 in 300.

18.—The manholes, ventilators, lampholes, and gullies to be built in accordance with the Local Surveyor's plans.

19.—The cross fall for roads should be, when practicable, half-an-inch to the foot, and for paved footpaths three-eighths of an inch to the foot.

20.—The fall from the kerbstone to the channel should not be less than three nor more than six inches, except in the case of crossings paved or formed for the use of foot passengers.

The preparation of Plans and Sections for Parliamentary deposit is dealt with in the following chapters. (Pages 239-308.)

CHAPTER XXI.

PARLIAMENTARY SURVEYING.

In the preparation of Plans and Sections for Parliament connected with projected schemes having reference to railways, tramways, docks, subways, water or gas supply, electric lighting and other works, the practice of surveying differs so much from ordinary chain or theodolite work, that a good expert surveyor of estates or parishes may often prove a slow surveyor for parliamentary work. The procedure is based upon certain regulations of both Houses of Parliament termed "Standing Orders," which have been framed with the object of giving due publicity to any application connected with a project to be laid before Parliament, in order that all parties who may be in any way affected by the proposed scheme may have a fair opportunity of stating their objections and acquire a *locus standi* before Committee to secure protection to their interests. In dealing with the various preliminary stages through which a private bill has to pass, the author will briefly indicate the purely legal proceedings with which it is necessary for a surveyor to be acquainted particularly as to prescribed dates, but the carrying out of which are for the most part beyond the province of a surveyor. The rules and practice of parliamentary work, however, demand an Engineer's knowledge of such regulations. As regards his own work the usual practice is to correct such portions of the Ordnance map to date as show the country which lies between the prescribed limits of deviation to be indicated upon the deposited plan, the principal object in the first instance being position and general direction rather than scrupulous accuracy of content. Every boundary, street, court, alley, dwelling house, yard and appurtenances within the limits of deviation must be shown upon the plan. In the case of a building estate he must notice if any pegs are in the ground, and if so, show the different allotments. In all cases descriptions should be borne in mind, and the plan enable the referencing, which will be more particularly described further on, and which is entrusted to the solicitor's department, to be clearly executed. Thus one field may have the same owner, but different tenants. At the same time no division need be shown on the plan which does not exist upon the ground. Difference of cultivation should be indicated by dotted lines. Roads that are metalled and all public footpaths must be shown; to ascertain whether they are public or not, the surveyor should see where they lead to. Thus in the case of a wood, it is well to walk round its boundary and note if there exists any gate, stile, or hurdle at the end of a footpath. Cut tracks leading as a *cul-de-sac* into a thick hedge need not be shown. It is not usual to show garden paths, but a footpath across a field leading to a church must be shown. All divisions that can be walked across are indicated by dotted lines. (See page 122.)

Prior to the publication of an Ordnance map (see pages 127 to 140) showing any particular country to be traversed, it was the custom to make tracings of selected portions of the parish or tithe map, then to piece such tracings together and to correct a copy of this combined tracing upon the ground. The method now adopted is to gum Ordnance maps together at their edges and to cut out long strips suitable for examination out of doors, containing sufficient of the surrounding roads and buildings to identify position outside the limits of deviation. A width of 18 inches is suitable for such strips, which should be rolled, not folded, and can be sent by post, if required, when not exceeding this length. The top of an old broomstick answers very well as a roller. When correcting the actual map in this way, a previous inspection of the area sheets published by the Ordnance Department may assist the surveyor in his observation of the leading characteristics of the property over which he travels, as these sheets show the state of cultivation of the different enclosures at the time the map was originally prepared. If a tracing of the Ordnance map be used for the field plan, it should be mounted on linen, but the scale should be drawn upon the tracing before it is mounted, so that it may shrink uniformly. The surveyor should particularly notice what land is common, or commonable land.

Courtesy to those who own or have the care of the land over which the surveyor and his men travel is one great element of success, otherwise considerable hindrance may be thrown in the way of progress. (See pages 189, 190.) A broad-size sketch-book, not too thick, is a useful companion, carried in an inside pocket of an overcoat, which the season of the year at which parliamentary work is usually executed compels a surveyor to wear. He also needs a tape and arrows and a chain for muddy roads, and he should take into consideration the short days of November, the likelihood of bad weather, and the limited time available prior to deposit of plans, in making his arrangements. The object of the sketch-book is to sketch buildings, yards, and enclosures, the divisions of which appear too close together upon the corrected Ordnance map to be distinct, so as to clearly show the reference numbers in case of dispute with the referencers. The surveyor must remember that unless his plan is self-explanatory it is practically useless, and hence a sketch should be made of all small and confined areas. An H. B. pencil is usually deemed the best for the purpose.

For the purposes of Standing Orders of each House, all private bills to which such regulations are applicable are divided into two classes, and are stated by the parliamentary agents to be of the first or of the second class, according to the subject to which they respectively relate:—

First Class.—1. Making, maintaining or altering a burial ground. 2. Enlarging or altering powers of charters and corporations. 3. Building, enlarging, repairing or maintaining a church or chapel. 4. Paving, lighting, watching, cleansing or improving a city or town. 5. Incorporating, regulating, or giving powers to a company. 6. County rate.

7. County or shire hall, court house. 8. Crown, church, or corporation property, or property held in trust for public or charitable purposes. 9. Ferry, where no work is to be executed. 10. Fishery, making, maintaining, or improving. 11. Gaol or house of correction. 12. Gas work. 13. Land, enclosing, draining, or improving. 14. Letters patent, confirming, prolonging, or transferring. 15. Constituting local court. 16. Erecting, improving, repairing, maintaining or regulating a market or market-place. 17. Police. 18. Maintaining or employing poor. 19. Poor rate. 20. Conferring powers to sue and be sued. 21. Payment of stipendiary magistrate, or any public officer. 22. Continuing or amending an Act passed for any of the purposes included in this or the second class, where no further work than such as was authorised by a former Act is proposed to be made. 23. Improvement charge unless proposed in connection with a second class work to be authorised by the proposed bill.

Second Class.—Making, maintaining, varying, extending or enlarging any aqueduct, archway, bridge, canal, cut, dock, or drainage where it is not provided in the bill that the cut shall not be more than eleven feet wide at the bottom, embankment for reclaiming land from the sea or any tidal river, ferry (where any work is to be executed), harbour, navigation, pier, port, railway, reservoir, sewer, street, public carriage road, subway, to be used for the conveyance of passengers, animals or goods in carriages or trucks drawn or propelled on rails, tunnel, "tramway," by which term as used in the Standing Orders is meant a tramway to be laid along a street or road, "tramroad," by which term as used in the Standing Orders is meant any tramway other than a tramway to be laid along a street or road, and last, though not least, a bill relating to waterworks.

In the House of Lords private bills are further distinguished as either local or personal, and bills for confirming a provisional order or provisional certificate are referred to as provisional order confirmation bills. The division into two classes applies to all bills, not being estate bills, which seek power with reference to any of the above subjects. In the House of Lords, arbitration in respect of the affairs of any company, corporation, or persons is added to bills of the first class.

A general list of plans that have been deposited is printed by the House, and can be obtained at the Private Bill Office of the House of Commons, upon application, early in the month of December.

Notices by Advertisement.—In all cases where application is intended to be made for leave to bring in a bill of either class, the Standing Orders of both Houses provide for notices to be given, stating the objects of the intended application and the powers intended to be applied for; and in the case of bills in respect to which plans are required to be deposited, such notices are ordered to contain a description of all the termini, together with the names of the parishes, townships, and townlands, together with extra-parochial places from, in, through, or into which the work is intended to be made, maintained, varied, extended, or enlarged, or in which any land or houses intended to be taken are situate, and where any common or commonable land is intended to be taken such notice shall contain the name of such common or commonable land (if any), with the name of any parish or township in which such land is situate, and the surveyor is called upon to furnish an estimate of the quantity of such common or commonable land proposed to be taken for insertion in the advertisement. In cases of bills for making a cemetery or burial ground, or for constructing gas works, or sewage works, or works for the manufacture or conversion of the residual products, or for a station for generating electric power, or for building a hospital for infectious disease, the notice shall specify the limits of land, in or upon which such cemetery or burial ground is intended to be made or such works constructed.

In the months of October and November, or either of them, immediately preceding the application to Parliament, the notices for a bill are published once in two successive weeks in some one and the same newspaper of the county in which such city, county of a city or borough, town, county of a town or district, or lands to which such bill relates shall be situate; or if there be no newspaper published therein, then in the newspaper of some county adjoining or near thereto ; and if such bill relate specially to any particular city, county of a city or borough, town or county of a town or urban district, in which any newspaper is published, the notices are published once in two successive weeks in one and the same newspaper published therein, with an interval between the two publications of not less than six clear days; or if such bill do not relate to any particular city, county of a city, town, county of a town or lands, such notices are published once in the London, Edinburgh, or Dublin Gazette only, as the case may be; and if such bill relate to lands situate in more than one county, such notices are inserted once in each of two successive weeks, in some newspaper or newspapers which are published in London at least six days in the week, or in Edinburgh or Dublin at least two days in the week, as the case may be, and in a newspaper of the county in which the principal office of the company or companies or other parties who are the promoters of any such bill is situate, and in a newspaper of every county in which any new works are proposed to be constructed, or in which any lands are intended to be taken, or in which any lands are situate in respect of which any new or further powers for the completion of works already authorised are intended to be applied for.

All advertisements are, nevertheless, in every case inserted once at least in the "London Gazette" if the place to which the intended application relates is situate in England or Wales, or in the "Edinburgh Gazette" if such place is situate in Scotland, or in the "Dublin Gazette" if such place is situate in Ireland.

The whole notice is to be included in one advertisement, which is to be headed with a short title descriptive of the undertaking or application. The advertisement for provisional orders and certificates is to be inserted once at least in each of two successive weeks in some one and the same newspaper published in the district affected by the proposed undertaking, or in the city, town, or place where the proposed works will be made; or if there be no such newspaper, then in some one and the same newspaper

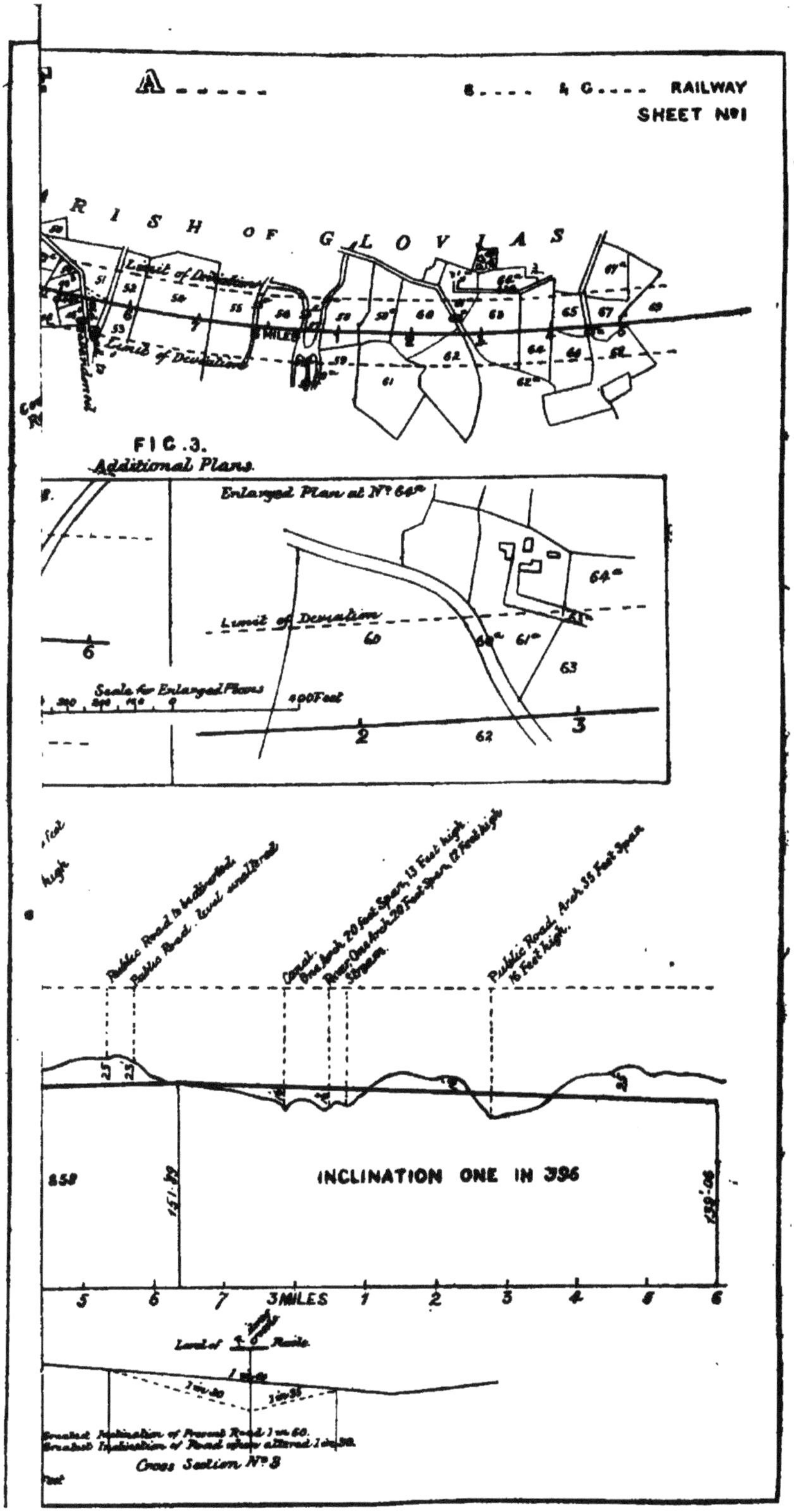

RAILWAY
SHEET Nº 1

PARISH OF GLOVLAS

FIG. 3.
Additional Plans.

Enlarged Plan at Nº 64ª

Limit of Deviation

Scale for Enlarged Plans

INCLINATION ONE IN 396

Cross Section Nº 3

published in the county in which such city, town, or place, or some part thereof, is situate; or if there be none, then in some one and the same newspaper published in some adjoining or neighbouring county.

A copy of the advertisement is, in all cases where plans are deposited, included by the Parliamentary Agent in the documents deposited by him on or before November 30th, as prescribed by the Standing Orders.

The principal part of this advertisement which affects the engineer or the surveyor is the description of the termini, which must be stated with sufficient exactness to enable any one interested to identify the spot within a few yards. The surveyor should show upon the plan the points from which the commencement and termination of the proposed line, in the case of a railway, have been measured, and record the total length in miles, furlongs, and chains. The zero on the proposed line would be described as " Commencement of Railway No. ," or, in the case of a junction, "Commencement of Railway No. and Junction with Railway No. ." The plans should be prepared with the utmost care, bearing in mind the ordeal to which they will be subjected. Attention should be specially directed, as stated above, to see that all enclosures existing on the ground appear on the plan, that all houses are inserted, of their correct shape and size, and gates across roads, which often divide a public road from an occupation one, shown. Often, too, a railway company will have a slip of land outside its thorn fence, of which no indication sometimes exists on the ground, the boundary of which should be shown by a dotted line and a separate number, but which is easily overlooked. The diagram, fig. 1 (pages 245-247), shows the form of a finished plan. The numbers in each enclosure refer to the book of reference, and are inserted by one of the reference clerks engaged by the solicitors to the undertaking, whose duty it is to state the names of the owners, lessees, and occupiers in his description of each enclosure. (Pages 255-260.)

It is the surveyor's work to see that all roads, fences, streams, buildings, county and parish boundaries, centre lines of proposed work, with ticks indicating miles and

L

furlongs, are accurately shown upon the tracing which is handed to the solicitor for reference. It will facilitate the solicitor's work if the buildings and roads are tinted upon the tracing he employs for this purpose. In examining the site of the proposed works with the aid of copies of the ordnance maps, it will be observed that sometimes a ditch is represented by a double line and sometimes by a single line. In order to make the field copy of the maps clear for reference when tracing for deposited plans, it will be well to write the letter f upon a line if a fence runs alongside a ditch, or to write the word "ditch" if only a ditch is represented on the field copy. The surveyor may use the proposed centre line on the map as a base line as much as possible, but in the case of new roads found to have been made since the plan was last revised he should take measurements from and along old fences in order to determine the position of the new road, and survey the fence from the centre line of the railway as a base.

In all cases where it is proposed to divert into any existing or intended cut, canal, reservoir, aqueduct, or navigation, or into any intended variation, extension, or enlargement thereof respectively, any water from any existing cut, canal, reservoir, aqueduct, or navigation, whether the water is to be abstracted directly or indirectly from any such cut, canal, reservoir, aqueduct, or navigation, or from any feeder thereof, and whether, under any agreement with the proprietors or otherwise, the notices by advertisement shall contain the name of every such last-mentioned cut, canal, reservoir, aqueduct, or navigation.

Under the Board of Trade regulations for provisional orders the advertisement and notices should state that every company, corporation, or person desirous of making any representation to the Board of Trade, or of bringing before them any objection respecting the application, may do so by letter addressed to the Assistant Secretary of the Railway Department of the Board of Trade, *on or before the 1st January* next ensuing; that copies of their objections must at the same time be sent to the promoters; and that in forwarding to the Board of Trade such objections, the objectors or their agents must state that a copy of the same has been sent to the promoters or their agents. The plans are prepared for deposit in the case of provisional orders under the acts relating to piers and harbours, tramways, gas and water, railways, and public health, in accordance with the rules and instructions issued by the Board of Trade and other public departments.

Gas and Water.

By the Gas and Water Works Facilities Act, 1870, it is provided that any company, companies, or person may obtain power under a provisional order confirmed by Parliament (1) to construct or to maintain and continue gas works and works connected therewith, or to manufacture and supply gas; also (2) to construct or to maintain and continue water works and works connected therewith, or to supply water in any district within which there is not an existing company, corporation, body of commissioners, or person empowered by Act of Parliament to construct such works and to supply water with the consent of such local and road authorities as exist in the district; and evidence of this consent is required to be produced at the time fixed for proving compliance with the Act and the Board of Trade rules. (3) To enable two or more companies or persons duly authorised to supply gas or water in any district or in adjoining districts to enter into agreements jointly to furnish such supply, or to amalgamate their undertakings. (4) To authorise two or more companies or persons supplying gas or water in any district or in adjoining districts to manufacture and supply gas or to supply water, and to enter into agreements jointly to furnish such supply and to amalgamate their undertakings; and such purposes, or any one or more of them, as the case may be, shall, for the purposes of the proposed Act, be deemed to be included in the term "gas undertaking" or "water undertaking," according as the same relate to the supply of gas or water; provided that any gas or water company empowered as aforesaid may apply for and avail themselves of the facilities of the proposed Act within their own districts respectively.

The promoters applying for a provisional order in

pursuance of the Gas and Waterworks Facilities Act shall, on or before the 1st November next, before their application, give notice in writing of their intention to make the same to every company, corporation, or person (if any) supplying gas (if the proposed application relates to gasworks) or water (if the proposed application relates to waterworks) within the district to which the proposed application refers.

The Board of Trade Rules, made under the abovementioned Gas and Water Facilities Act, 1870 (33 and 34 Vict. cap. 70), with respect to Provisional Orders are printed in a publication of eight pages of foolscap which give a description of cases within the Act by which the necessary consents required for Provisional Orders may be obtained ; also the nature of the advertisements, and of the notices to be given in the months of October or November, and in December, the deposits to be made on or before November 30, and on or before December 23, the regulations as to proof of compliance with the Act and with the Board of Trade rules, also as to deposit and publication of the advertisement, together with requirements for any prolongation of time for the commencement or the completion of the works, and a model of a plan of gasworks which has to be given to a scale of not less than 100 feet to one inch. The publication shows by whom Provisional Orders may be obtained, and is issued by the Board of Trade, Railway Department, dated August, 1886. It prescribes that all memorials, objections, and other documents should be furnished upon paper of foolscap size, and that promoters who desire to be incorporated must register themselves under the Companies Act, 1862.

The granting of Provisional Orders authorising the supply of electricity and giving the Board of Trade the authority to make from time to time rules in relation to applications for licences or Provisional Orders, for the purpose of facilitating and regulating the supply of electricity for lighting and other purposes in Great Britain and Ireland, is prescribed under the Electric Lighting Act, 1882 (45 and 46 Vict. cap. 56). This Act also provided that within six months after the expiration of a period of 21 years, or as otherwise provided by the special Act, the local authority may acquire from the promoters or subsequent owners the undertaking by purchase, upon conditions furnished by the clauses of this Act. In 1888 another Act was passed, and in 1899 a further Act which consolidates all the necessary clauses. The 1882 Act costs 1s., the 1888 Act costs 1s. 2d., and the 1899 Act costs 1s., and copies can be purchased at the Queen's Printers.

TRAMWAYS AND SUBWAYS.

The notice by advertisement shall state what power it is intended to employ for moving carriages or trucks upon a street tramway or upon a tramroad, and in the case of bills for constructing *a subway*, the notice shall specify the gauge to be adopted and the motive power to be employed.

Street Notices.—In either the month of October or November immediately preceding the application for any bill or provisional order for laying down a tramway or constructing an underground railway or a subway, when such bill contains powers authorising any alteration or disturbance of the street or road surface, notice thereof shall be posted for 14 consecutive days in any street or road along or under which it is proposed to lay the tramway or construct the subway, and placed in such manner as the authority having the control of such street or road shall direct; but if after application to such authority no such direction shall be given, then the notice shall be placed in some conspicuous position, such as the lamp posts of the street or road. The notice must clearly state the place or places at which the plan of such tramway, railway, or subway will be deposited, for the information of the public.

Consents in Case of Tramway Bills.—In cases of bills to authorise the laying down of a tramway along any public highway, the promoters shall obtain the consent of the local authority of the district or districts through which it is proposed to construct such tramway, and where in any district there is a road authority distinct from the local authority, the consent of such road authority shall also be necessary in any case where power is sought to break up any road subject to the jurisdiction of such road authority. For the purposes of this order, the local and road authorities in England and Scotland shall be the local and road authorities mentioned in section 3 and schedule A of "The Tramways Act, 1870;" and in Ireland shall be the grand jury of the county in respect to any highway or portion of highway within the jurisdiction of such grand jury; and in respect to highways wholly or partly within any city, borough, town corporate, or other place or district in which the public roads are not under the control of the grand jury of the county, shall be the respective local and road authorities of such city, borough, town corporate, or other place or district mentioned in section 38 of "The Tramways (Ireland) Act, 1860." Provided that where it is proposed to lay down a continuous line of tramway in two or more districts, and any local or road authority having jurisdiction in any such districts does not consent thereto, the consents of the local and road authority or the local and road authorities having jurisdiction over two-thirds in length of the highways along which such tramways are proposed to be laid shall be deemed to be sufficient.

The evidence of the consent of the local and road authorities and all other proofs of compliance with the Act and the Board of Trade rules must be completed on or before the 15th February, and any clause which a local authority, railway company, canal company, or others may desire to have inserted in the order must be submitted to the Board of Trade by that date.

The Board of Trade have published a pamphlet of 18 pages accompanied with a diagram (specimen) of a form of plan for proposed tramways to a scale of not less than two inches to one mile, as required to be deposited at the Board of Trade, with the plans, on or before the 30th of November. This publication, a printed copy of which may be obtained at the Queen's Printers, price fourpence, shows (1) by whom a Provisional Order may be obtained under The Tramways Act, 1870 (33 and 34 Vict. cap. 78); (2) the required consents to such Order; (3) the advertisements and notices needed to be given in October or November and in December; (4) the deposits to be made on or before November 30; (5) the deposits to be made on or before

December 23; (6) the form of the draft Provisional Order;
(7) the proofs of compliance with the Act and rules;
(8) regulations as to the deposit and advertisement of the
Order as made, and of any amended plans and sections;
(9) deposits of money in the Chancery division, penalty for
non-completion of tramways and release of the deposit;
(10) rules as to opening of any new tramway, and finally
(11) prolongation of time for the commencement or completion of the works. As in the case of Gas and Water
Orders, all memorials, objections, and other documents
have to be furnished upon paper of foolscap size, and
promoters who desire to be incorporated must register
themselves under the Companies Act, 1862.

In The Tramways Act, 1870, and other Acts of
Parliament above alluded to, the name of the London
County Council must be substituted for the Metropolitan
Board of Works, Urban District Council for Local Board,
and Rural District Council for Vestry, Select Vestry, or
other body of persons acting by virtue of any Act of
Parliament, prescription, custom, or otherwise, as or instead
of a previously constituted Vestry or Select Vestry.

Pier or Harbour.

In October, 1884, the Harbour Department of the Board
of Trade issued certain regulations as to Provisional Orders,
and also provided for any application for a Provisional
Order relative to a Pier or Harbour, under "The General
Pier and Harbour Act, 1861," and under "The Harbour
Transfer Act, 1862." The latest issue is dated October,
1899. Copies of this memorandum should be obtained by
Surveyors to Harbour Authorities and those interested in
works situated on tidal lands. The publication now consists of
20 printed pages of foolscap, and can be purchased at Messrs.
Eyre & Spottiswoode's, East Harding Street, London.

General Plan.

In the case of all bills of the first class, the plan must
be such as will describe the lands intended to be taken.
In the case of all bills of the second class, the plan fully
describes the line or situation of the whole of the work (no

alternative line or work being in any case permitted), and the lands in or through which it is to be made, maintained, varied, extended, or enlarged, or through which any communication to or from the work shall be made; and where it is the intention of the promoters to apply for powers to make any lateral deviation from the line of the proposed work, the limits of such deviation shall be defined upon the plan, and all lands included within such limits shall be marked thereon.

The limits of deviation of the centre line are imaginary lines of boundary between which the work may be deviated laterally if it is required to shift the centre line during construction. Where in the case of a projected railway, any special property either wholly or partly situated within four-and-a-half chains upon each side of the centre line is decided not to be interfered with, the limits of deviation may be contracted by being drawn round the edge, but within the boundary of the enclosure referenced, so as to show upon the proposed railway side of the boundary of this property. If drawn over any part of the property, the space enclosed by the line of the limit of deviation must be numbered however small or inappreciably narrow that space may be. To ensure the fences being in every case drawn up to and correctly between the limits of deviation, it is well to provide that upon the revised tracing the fences shall in all cases be drawn, so as to just cross the limits of deviation. Further extension beyond the limits of deviation is not necessary. The surveyor adjusts the position of fences, &c., on his original plan to suit the dimensions taken in the field, so that the opposition measuring from these points for the section may find the levels agree. In the addition of new buildings and of new fences to the plan during the process of examination and correction, the surveyor must be careful of direction, as he not only completes a plan, but is providing thereby on the plan points for the opposition to use in setting out for checking the work. Therefore, everything within the limits of deviation ought to measure correctly off the centre line on the plan.

The Book of Reference.—The Draft Book of Reference is

divided into six columns, as follows, but in the copies to be deposited with the clerks of the peace, &c., the sixth column, intended for office information only, must be omitted.

It is headed as follows:—

The Parish of   in the County of

Number on Plan.	Description of Property.	Owners or Reputed Owners.	Lessees or Reputed Lessees.	Occupiers.	Observations.
1	2	3	4	5	6

The first column should contain the number of each property through which the line is laid down or which lies within the limits of deviation marked upon the plan; but where the limit of deviation crosses an enclosure, the number may be written within the plot outside the limit of deviation for sake of clearness. Each separate property should be numbered on the plan, and a corresponding number placed in the book of reference. Any collection of buildings and ground within the curtilage of a building, belonging to the same owner and in the same occupation, such as a "farmhouse, farmyard, barn, and outbuildings," may be described under one number. Where the occupations are different, although the same owner, there should be two separate numbers. The numbering should recommence in every parish, and therefore should not be continuous throughout the plan. When a river or brook divides the parishes the boundary line is generally in the mid-stream, and the description should be given for each parish. Where, as in the case of an island in the centre of a river through which the parish boundary runs, the exact division is ambiguous, the land may be described on the plan as being "in the parish of   or in the parish of   one or both of them," and referenced accordingly. The reference takers should be instructed to begin numbering at the same end of each parish, so that the numbers may run in one direction from the commencement of the line of the proposed work to its terminus, as described in the Gazette notice. If it is necessary to interpose a number, a duplicate number may be added, thus, 2A, but such repetition of numbers should wherever possible be avoided. All properties of which any portion is included within the limits of deviation, as drawn on the plan, should be described and inserted in the book of reference.

The second column should contain a description of the property indicated by the number, such as "manufactory, offices, warehouses, and premises," or as the case may be, "pasture field, arable field, park, plantation, wood, orchard, garden, house, yard, shed, house and outbuildings, turnpike road, public highway, occupation road, river, brook, common, waste, &c.," or by whatever other title, not being a local name, the same may be properly designated. In the case of fields through which a public footpath or an occupation road runs, not divided from the field by a fence, it will be better to describe it as "pasture field and footpath or occupation road," or "arable field and footpath or occupation road," as the case may be. If the occupation road is fenced off, it should be separately numbered. In all cases the descriptions should be full and complete.

The third column should contain the names of the owners or reputed owners of the property referred to by the particular number. The names both of the persons in the receipt and in the beneficial enjoyment of the rents, whether tenants in fee or for life only, should be inserted. It will be advisable, in describing trustees, to adopt an uniform method of description, as "A. B. and C. his wife, and their trustees, D. E. and F. G.," or "A. B. and C. D., the devisees in trust under the will of E. F." Where mortgagees are in possession, their names should also be inserted as well as that of the mortgagor; the like with assignees of a bankrupt's estate. If the late owner be very recently dead and the name of his heir, devisee, or devisee in trust, cannot be learnt, insert "The representatives of the late A. B. deceased."

In cases of glebe land, the name of the incumbent, as

well as that of the patron of the living, should be inserted. Where lands belong to infants or lunatics, their names should be inserted with the names of their guardians, or committees. The "lord of the manor" should be inserted as the owner of all waste lands and commons. In cases of corporate owners, insert them under their corporate names. In the case of charity or other public trustees not incorporated, the individuals should be named, and it will be convenient to name their clerks also. If on enquiry of their clerk he declines to give the names of the individuals, and they cannot be ascertained from any other source, the ordinary title of the trust, with the name of the clerk or receiver, should be given. The "trustees" of a turnpike road should be inserted as the owners of it, and should be named as "The trustees of the turnpike road, leading from A to B," or "The trustees of the   turnpike roads," as the case may be, learning the name of the clerk.

The surveyors of the parish or township should be inserted as the owners of public highways, and their names should be ascertained and specified. It will frequently be found that there are several highway districts in one parish, and in such cases the names of the surveyors and of the district should be specified as to each road.

Public roads should have a separate number in each parish where they appear on the plan, and all roads which are fenced off from adjoining land. Footpaths if repaired by the parish or if fenced off should be numbered. Navigable streams and mill streams should be separately numbered.

With regard to "private" or "occupation roads," where they do not belong to one particular person, the owners of the lands numbered on the plan immediately adjoining on either side of the point crossed by the railway, and also within the limits of deviation, should be named as owners. The same may be said of rivers and streams (not navigable). If the stream is dammed up, and used to turn a mill, the name of the mill-owner should also be added.

The names of the several lords of manors should in all cases be inserted.

The fourth column should contain the names of such

parties as have an interest under a lease, whether as lessees or under lessees. Where the occupier and lessee are one and the same person, the name should appear both in the lessees' and occupiers' column.

The fifth column will be devoted to the *occupiers* of the several properties, indicated by the numbers. The names of these can, of course, be easily ascertained. The only difficulty appears with regard to the occupiers of commons. The commoners here certainly have an interest, but they are generally too numerous to be inserted in the book of reference. They generally, however, have a person called the hay-warden or field reeve, cattle driver, or bailiff, who is appointed to look after the stock on the common, and his name should be ascertained and inserted. The same may be said of open field lands or Lammas lands, where certain persons have a right of turning in their cattle at particular periods of the year, and the surveyor gives attention to the position of the Lammas and commonable land that is included by the referencer, so that the surveyor may confirm its description and estimate its area. It is unnecessary to insert the names of occupiers of private or occupation roads or of streams, unless used to turn a mill, where dammed up as before mentioned, in which case the name of the occupier of the mill should be inserted. In passing through towns lighted with gas or electricity or supplied with water, established under Act of Parliament, the corporate bodies should be inserted as the occupiers of the streets where the pipes or mains are laid, and H.M. Postmaster-General must not be forgotten when referencing a railway where telegraph wires are included.

The sixth column will be directed to observations for the information of the promoters, and should, in all cases, contain the names of the persons who have supplied the information inserted in the previous columns, together with all such other remarks as may tend to facilitate checking the information already obtained, or the obtaining of further information. It should also contain the names and residences of all agents, &c., who may be referred to as possessing information respecting the particular properties, and, generally, such other observations as may appear likely to be useful. If the owners or lessees are not resident near the spot, particular care should be taken to learn accurately their present as well as usual place of abode; and where parties are absent from the United Kingdom, the name and residence of the agent who receives the rents of the property; and these addresses must be inserted in this column with some denoting mark when the address given refers to the address of the lessee, as otherwise it will be assumed to refer only to the address of the owner.

To guide the referencers it is usual to tint the roads and streams as well as to tint with a red colour dwelling-houses and with a black colour outbuildings, upon the tracing or map supplied to them, whereas upon the deposited plan the numbers alone serve to show whether any enclosure is a house or a field.

The plan, or a mounted tracing of the plan, having been put into the hands of the different gentlemen who are to get up the book of reference, they will do well to procure the assistant overseer, collector of parish rates, or other person having the means of supplying the required information, to accompany them through his parish or township, paying him, of course, for his time. After walking with him over the properties described in the plan, and setting down, from his information, the whole particulars above required, they must make a point of seeing personally every occupier, and ascertaining from him if all those particulars are correct. The occupier must be questioned as to each particular comprised in the several columns, and should be requested to accompany the reference takers over the property in his occupation. And it must be carefully borne in mind that before the examiner on standing orders no excuse will be allowed for mistakes or omissions, unless the occupier has been personally seen, he being the best authority; and the promoters being required to show that they have used their best endeavours to supply correct information. The tenant should be also questioned as to whether or not he holds his land upon a lease; also, whether he pays his rent to the lessor named in his lease or to any other person, and to whom; also, whether any part of his land is extra-parochial.

In case of house property, application should first be

made to the occupier for the name of the lessee, and afterwards to the lessees for the name of the ground landlord. The christian names of all parties must be ascertained, and the addresses placed in the column of observations.

The boundaries of the parishes should also be ascertained and clearly defined upon the plan, and the reference takers must satisfy themselves that these are correctly laid down, and inquiry should be made of the occupiers on this point. There are, in many places, parish maps showing each field, with the name of the owner and occupier; these generally cannot be relied upon, but may sometimes be advantageously referred to, particularly with respect to the parish boundaries.

The persons engaged should, in the evening, fair copy the work of each day and transmit such fair copy as instructed. Care must be taken that the numbers on the tracings are not obliterated. As from the delay of the surveys and other causes the utmost exertion and diligence will be necessary to complete the book of reference in time for deposit, the following suggestions are made : The parties taking the reference should station themselves at an inn nearest to the part of the line of work upon which they are engaged and remove their quarters as they proceed. They should be upon the ground early every morning and remain seeing the occupiers in the neighbourhood until dusk, and after that time they should see any more distant parties from whom it may be proper or desirable to obtain information, so that, as far as possible, each day's work may be completed while it is fresh in the recollection of the persons engaged. Inconvenience invariably arises from the practice of putting off these inquiries from day to day. About once a week they should devote a day, or such part of a day as may be necessary, to seeing any occupiers or parties whose residence at a distance renders it inconvenient to see them upon the days on which they were going over the line. This advice as to the apportionment of time must, of course, be subject to any special circumstance. Whenever an inquiry has to be made from a party at a distance, to whom the person making the reference cannot go, a note of the inquiry to be made should be written on the fair copy of the reference. The words "saw A.B." "saw C.D.," &c., should be written against the name of the occupier whenever it occurs. Where there are two persons of the same christian and surname, owners or occupiers in the same parish, they should be distinguished by their addresses, or otherwise. The plan should be numbered either in pencil or ink while they are upon the ground, and should not be left to the memory to supply the numbers in the evening. Any properties which have not been numbered, and are included in the limits of deviation, must be numbered with the same number as on one of the adjoining properties, adding the letters *a, b, c,* &c., as may be required. In no case should the word "ditto" be used in the book of reference or the drafts, but the descriptions and names repeated in every instance against every number. Finally the Gazette notice should be examined, in order to ascertain that all the parishes and townships through which the line passes are included in it.

Instructions for the preparation of a book of reference containing many of the above valuable hints as to the manner of obtaining the necessary information as to the course of proceeding, are published in a printed form by Messrs. Vacher & Sons, of Westminster.

Plan.—Every plan required to be deposited must under the requirements of the Standing Orders be drawn to a scale of not less than four inches to a mile, or 20 chains (1,320 feet) to an inch, but unless the whole plan is upon a minimum scale of six chains or 396 feet (practically 400 feet) to the inch = $\frac{1}{4800}$, (six chains to one inch = $\frac{1}{4752}$), an enlarged plan must be added of every building, yard, courtyard or land within the curtilage of any building, and of every enclosed area, whether open or covered, or of any ground cultivated as a garden, either in the line of the proposed work or included within the limits of the said deviation, upon a scale of not less than a quarter of an inch to every one hundred feet; all the reference numbers must likewise be repeated as near as possible on the reduced plan, and care taken that the enclosures shown upon the enlarged plan, as indicated in fig. 3, not only bear the same reference numbers as those upon the smaller scale plan, but that the enlarged plan shows the centre line and limits of deviation, though not necessarily the furlong ticks, as these are indicated on the reduced plan. (Pages 245-247.)

The choice will naturally lie between the plan of 6 inches to the mile, which economises much in cost of lithography, for a long line, but necessitates enlarged plans, and a plan drawn to the scale of the parish survey of 25 inches to the mile. This latter is, in the author's opinion, the best to adopt for all suburban work, or where a large number of enlargements would be required. Larger scales even than this are sometimes adopted for metropolitan work, 88 feet to the inch being frequently used; but where, on the other hand, enlarged plans are but seldom needed, the six-inch scale will be the best to adopt.

The plan and book of reference must not only accurately describe the ground, but must, as before stated, agree with one another. Thus, with regard to tram lines, if the public road is referenced as "public road and tramway," it is desirable to indicate such tramway upon the plan. This would be almost impossible upon a 6 inch to the mile scale, but where a larger scale is employed they should be inserted. It is obvious that in many cases the reference numbers cannot be written so distinctly upon a reduced as upon an enlarged plan, and hence the value of adopting a scale of, say, 6 chains to the inch, as it saves the necessity of enlarged plans. Where a line is not likely to be much opposed, a copy of the ordnance map to the scale of 25 inches to the mile may be recommended for all suburban work, or where a large number of enlargements would be needed, while it must be remembered that the map to the scale of 6 inches to the mile renders the referencing more difficult to check.

WATERWAY.

In all cases where it is proposed to make, vary, extend, or enlarge any cut, canal, reservoir, aqueduct, or navigation, the plan shall describe the brooks and streams to be directly diverted into such intended cut, canal, reservoir, aqueduct, or navigation, or into any variation, extension, or enlargement thereof respectively, for supplying the same with water.

SUBWAY.

In the case of applications for constructing a subway, the plan and section shall indicate the height and width of the proposed subway and the nature of the approaches by which it is proposed to afford access to such subway.

TRAMWAY.

In cases of application for laying down a street tramway, both the notice by advertisement and the plan shall indicate whether it is proposed to lay such tramway along the centre of any street, and if not along the centre, then on which side of, and at what distance from an imaginary line drawn along the centre of such street, and whether or not, and if so, at what point or points it is proposed to lay such tramway, so that for a distance of thirty feet or upwards a less space than nine feet six inches, or if it is intended to run thereon carriages or trucks adapted for use upon railways, a less space than ten feet six inches shall intervene between the outside of the footpath on either side of the road and the nearest rail of the tramway.

All places where for a distance of thirty feet and upwards there will be a less space than nine feet six inches between the outside of the footpath on either side of the road, and the nearest rail of the tramway shall be indicated by a thick dotted line on the plan on the side or sides of the line of tramway where such narrow places occur, as well as noted on the plans, and the width of the road at those places should also be marked on the plans.

All lengths shall be stated on the plan and section in miles, furlongs, chains, and decimals of a chain. The distances in miles and furlongs from one of the termini of each tramway shall be marked on the plan and section. Each double portion of tramway, whether a passing-place or otherwise, shall be indicated by a double line.

The total length of the road upon which each tramway is to be laid shall be stated, *i.e.*, the length of route of each tramway.

The length of each double and single portion of such tramway, and the total length of such double and single portions respectively shall also be stated.

In the case of double lines (including passing-places), the distance between the centre lines of each line of tramway shall be marked on the plan. This distance must in all cases be sufficient to leave at least fifteen inches between the sides of the widest carriages and engines to be used on the tramways when passing one another. The above regulations also apply in the case of a tramroad when worked along a road surface.

Railways.

Maximum limits of deviation for railways.—In towns, 10 yards each side of centre line. In the country, 100 yards, about $4\frac{1}{2}$ chains ($4\frac{1}{2} \times 22$ yards = 99 yards).

Maximum deviations of curves in construction for railways.—Curves upward of $\frac{1}{2}$ a mile radius, may be sharpened to $\frac{1}{4}$ mile radius. Curves of less than $\frac{1}{4}$ mile radius, may not be sharpened. Curves of less than 80 chains, or 1 mile radius, must have the radius written against the line of curve in furlongs and chains. The beginning and end of the curve should be indicated by a cross thus —✕— upon the centre line. These are termed tangent points, and must be shown on the plan.

The centre line of a proposed railway is shown upon the plan by a strong black line, upon which the miles and furlongs are indicated by short cross black lines upon one side of the centre line, called furlong ticks and mile points. The distances of the mile and furlong points are also indicated upon the plan in their correct position.

Where tunnelling as a substitute for open cutting is intended, the centre line is shown by a dotted line on the plan, and no work should be shown as tunnelling in the making of which it will be necessary to cut through or remove the surface soil.

If it be intended to divert, widen, or narrow any public

carriage road, navigable river, canal or railway, the course of such diversion and the extent of such widening or narrowing should be marked upon the plan.

When a railway is intended to form a junction with an existing or authorised line of railway, the course of such existing or authorised line of railway should be shown on the deposited plan, on the same scale as the scale of the general plan. In the case of two proposed railways No. 1 and No. 2 making a junction one with the other, the surveyor would write on the plan of railway No. 1 at the proper point, "junction of railway No. 2," and upon railway No. 2 he he would write "termination of railway No. 2 and junction with railway No. 1." The gradient of an existing or authorised line of railway to which the new line is proposed to be connected is also shown on the deposited section, and in connection therewith, and on the same scale as the general section, for a distance of 800 yards (approximately 40 chains of 66 feet or $\frac{1}{2}$ mile section) upon either side of the point of the proposed junction. The point of junction upon a plan needs careful determination and description.

Lands for extraordinary purposes.—The company may take by agreement for the extraordinary purposes mentioned in the Railways Clauses Consolidation Act, 1845, in connection with the new railways and the deviation and widenings, any quantity of land not exceeding 10 acres.

Light railways.—There are three publications, (1) Statutory Rules and Orders, 1896 (Paper No. 787, 8 pages, price $\frac{1}{2}$d.), issued by the Board of Trade for regulating the procedure before the Light Railway Commissioners, where a scheme for a (so-called) light railway has been matured and it is intended to make a formal application for a Provisional Order, (2) Statutory Rules and Orders, 1898 (Paper No. 705, 10 pages, price 1d.), containing modifications of the preceding rules and order, and (3) a list of Rules, dated May, 1898, issued by the Board of Trade, with the concurrence of the Lord Chancellor, pursuant to the thirteenth section of the Light Railways Act, 1896 (Paper No. $\frac{496}{\text{L. 13,}}$ 5 pages, price $\frac{1}{2}$d.). The two former publications deal with the notice by advertisement and the required contents of that notice, the deposit thereof, together with plans and sections at the offices of the local authorities and Government departments, regulations affecting the preparation of the plans and sections and book of reference, notices to owners, lessees and others, estimate of the various branches of each light railway, and general summary of total cost, particulars of a fee of £50 to be paid by the promoters to the Board of Trade, and a schedule furnishing the form of notice to the landowners and others. The last named of these three publications deals mainly with the costs to be allowed and taxation as against a light railway company, for the charges of a claimant in an arbitration under the Act. The proposed gauge and motive power of the railway must be stated in the notice, and also the place where copies of the draft order can be obtained on payment of not exceeding one shilling a copy. The above printed rules and orders can be purchased at Messrs. Eyre & Spottiswoode, the Queen's printers, and the Light Railway Commissioners undertake to give every further facility in their power for considering and maturing proposals for the construction of light railways to be submitted to them. The present address of the Light Railway Commission is 54, Parliament Street, and communications should be on foolscap paper written upon one side only. It is mainly the prescribed notice to frontages, as stated on pages 286 and 293, that has to be considered in deciding whether a scheme shall be proposed as a light railway, tramroad, or tramway in country districts.

General Summary (Plan).

The following data should be read through in the completion of every railway plan in order to ascertain that the descriptions are properly shown upon the plans to be deposited.

(a) Title to be given.
(b) Centre line to be shown.

    1.—Commencement of railway.
    2.—Junction with railways.
    3.—Mile and furlong points.
    4.—Termination of railway—length in miles, furlongs, and chains.

PARLIAMENTARY SURVEYING. 268

5.—Junction of railway at end of line.
6.—Ditto with branches.
7.—Plan showing 800 yards each way, existing railway.
8.—Radii of curves under one mile, in furlongs and chains.
9.—County and parish and townships.
10.—Sheet No.
11.—Limits of deviation (to be closed).
12.—Reference numbers.
13.—Road diversions or stopped up to be shown.
14.—Tunnel and viaduct lengths.
15.—Tunnel dotted.
16.—Cut and cover, not tunnel.
17.—Viaduct.
18.—Scales.
19.—Board of Trade copy, tidal waters blue.
20.—Parish and county and township boundaries.
21.—Cut lines. (See plan agrees.)
22.—Letter plan where cross sections require same.
23.—Enlargements. (See Enlargements A, &c.)
24.—Authorised railways to be shown, centre line.
25.—Show by dotted line existing tunnels, and describe.
26.—Datum point marked on plan.

Section.—An accurate plan is important in order to make an accurate section. (See page 163.) The section connected with a Parliamentary Plan is seldom chained from end to end, but its position in open country is measured from well-defined points on the plan in the nearest fences within the limits of deviation, and in towns from the corners of buildings shown on the plan, so as to work in straight portions and curves correctly. These measurements should be scaled off the original corrected map in the office, and figured upon the tracing that is to be used for setting out in the field. Fig. 2 shows a portion of a parliamentary section, in which it will be observed that according to the Standing Orders the section is drawn to the same horizontal scale as the plan. (See pages 245-247.)

The section shall be drawn to a vertical scale of not less than one inch to every 100 feet, and show the surface of the ground marked on the plan, the intended level of the proposed work, the height of every embankment and the depth of every cutting, and a datum horizontal line, which shall be the same throughout the whole length of the work, or any branch thereof respectively, and is referred to some fixed well defined point (stated in writing on the section), near some portion of such work, and in the case of a canal, cut, navigation, public carriage road or railway, near either of the termini. It is therefore important to see that the datum point is clearly shown upon the plan and properly described upon the section. Where for convenience of checking the section, the Ordnance levels have been followed, and an Ordnance bench mark has been adopted as the fixed datum point, it must be referred to as a mark cut thus ⏉ upon the defined position as stated in fig. 2 without describing it as an Ordnance mark. The point should be indicated upon the plan in its exact position, but it is well not to trust to the stated level of a single Ordnance bench mark in determining the level of a starting point without checking it with other Ordnance bench marks near at hand. The mileage and furlong ticks must be shown to correspond with the plan. Where the level of any road is intended to be altered in making any public work, the ascent of any turnpike road, or of any road if in Ireland so defined in the Railway Clauses Consolidation Act, 1845, shall not be more than one foot in 30 feet, and of any other public carriage road not more than one foot in 20 feet, and a good and sufficient fence of from four feet high at the least shall be made on each side of every bridge which shall be erected. Any deviation from this regulation needs a report thereon from some officer of the Board of Trade, which shall be considered when the bill comes before a Committee of the House.

RAILWAY.

Plans and Sections.—The following regulations for railways must be observed in the case of all public roads that are crossed upon the level. All other public roads are noted upon the section, and the method of dealing with them particularly described.

PARLIAMENTARY SURVEYING.

	Public road. ft. in.	Public road. ft. in.	Occupation road. ft. in.
Clear width of under bridge or approach	35 0	25 0	12 0
Clear height of under bridge for a width of 12 ft.	16 0	—	—
Ditto for a width of 10 ft.	—	15 0	—
Ditto for a width of 9 ft.	—	—	14 0
Ditto at springing	12 0	12 0	
Over bridge height of parapets above the level of the rails	4 6	4 6	4 6
Under bridge, height of parapets above the level of the rails	4 6	4 6	4 6
Least thickness of parapets 18 in. in the case of under bridges and viaducts			
Approaches inclination	1 in 30	1 in 20	1 in 16
Ditto height of fencing	3 0	3 0	3 0

Widths of centres allowed in calculation.—See page 319.

Maximum deviations of level.—In towns, 2 feet are allowed. In the country, 5 feet. *Maximum deviations of gradient for railways.*—Gradients flatter than 1 in 100, deviation 10 feet per mile steeper is allowed, but when steeper than this gradient, 3 feet per mile steeper is the limit.

In every section of a railway, the line of the railway marked thereon shall correspond with the upper surface of the rails, and the horizontal distances on the datum line are marked in miles and furlongs, to correspond exactly in length with those on the plan, and also with the horizontal scale drawn upon the plan. Hence, in fixing the position of points for plotting the levels off the plan, it is well to measure from the furlong ticks, and to see that the relative position of such points correspond accurately upon both the plan and section.

Either the horizontal and vertical scales must be drawn upon each sheet or upon the first sheet, giving scales only as a separate sheet and applicable for all sheets. A vertical measure from the datum line to the line of the railway should be legibly written in feet and inches, or decimal parts of a foot, at the commencement and termination of the railway, and also at each change of the gradient or inclination thereof; and the proportion or rate of inclination between every two consecutive vertical measures should also be indicated. These vertical heights must scale accurately to the vertical scale drawn upon the plan, and the gradients require therefore to be very accurately calculated from the vertical heights written upon the section and from the scaled horizontal distances. The rail level is described as "line corresponding with upper surface of rails." In the case of a tramway, the gradients of the road on which each tramway is proposed to be laid should be indicated upon the section.

Wherever the line of the railway is intended to cross any public carriage road, navigable river, canal, or railway, the height of the railway over or depth under the surface thereof, and the height and span measured on the square of every arch of all bridges and viaducts by which the railway will be carried over the same, should be marked in figures at every crossing thereof. The depths of all cuttings and embankments at their deepest points when over 5 feet above or below rail level should be figured on the section to the surface level. The datum for railway No. 2 may be described, if desired, as " datum same as for railway No. ..." and so also for other proposed lines.

Where the railway is proposed to be carried across any such public carriage road or railway on the level thereof, such crossing shall be so described on the section; and if such level will be unaltered, it shall be stated thus : " Road level unaltered," and the span and height of arch stated according to the preceding table (page 270). There is no occasion to burden the section by unnecessary work for the lithographers, such as writing up the words "street" or "road" to an occupation road. It is not necessary to write "public road" upon the section. This is no more needed than to describe a field as "arable" in the book of reference. The very fact of noticing the road on the section proves it to be public, but nevertheless it is frequently so described. The diameter of all culverts must be shown. If any alteration be intended in the water-level of any canal, or in the level or rate of inclination of any public carriage road, or railway, which is proposed to be crossed by the line of railway, then the same shall be stated on the section, and each alteration is numbered; and cross sections, in reference to their numerical order, on a horizontal scale of not less than 1 inch to every 330 feet, and on a vertical scale of not less than 1 inch to every 40 feet are added, which show the present surface of such road, canal, or railway, and the intended surface thereof when altered; and the greatest of the present and intended rates of inclination of the portion of such road or railway intended to be altered should also be marked in figures thereon.

Where any important road or public carriage road is crossed on the level, a cross section of such road should also be taken whether the level is required to be altered or not, and all such cross sections should extend for 200 yards on each side of the centre line of the railway. A length of 10 chains (220 yards) is generally the distance which is measured on either side of the proposed centre line of the railway for this purpose. In such cross sections the surveyor first fixes the height by scale, and writes on in feet and inches the greatest depth of cutting or height of bank connected therewith. It is necessary to remember that the cross section should show, in figures as well as by measurement, the present level and the intended level of the road; but the greatest inclination of the present road is not to be taken at the steepest point within the 200 yards upon each side of the line, but between the points within which the road is to be altered. Or if no alteration is required the description will be "public road level unaltered and crossed on the level," height written on as 0·00, or if altered, then "public road to be raised   feet, see C. S. No.   ."

In the case of railways, the Act, when necessary, may provide clauses as follows :—

Power to divert roads as shown on deposited plans.— The company may also obtain power by a clause in their Act to divert, alter, and stop up in the manner shown upon the deposited plan and section any roads delineated on the said plan, and described in the deposited book of reference, and when and as in each case the new portion of any road is made to the satisfaction of two Justices of the Peace, and is open for public use, may stop up and cause to be discontinued as a road so much of the existing road as will be rendered unnecessary by the new portion of road, and when and so soon as each of the said roads is so stopped up, all rights of way over the same shall cease, and the company may, subject to the provisions of the Railways Clauses Consolidation Act, 1845, with respect to mines lying under or near to the railway, appropriate and use for the purposes of their undertaking the site of the road stopped up as far as the same is bounded on both sides by lands of the company.

Referring to the inclination of roads.—In altering for the purposes of the widening No.   at   by this Act authorised, the roads numbered on the deposited plans   and   in the parish of   the company may make the same of any inclinations not steeper than one in sixteen.

Width of a certain roadway.—The company may make the roadway over the bridge, by which the road numbered on the deposited plans   in the parish of   will be carried over the widening No.   at   of such width between the fences thereof as the company think fit, not being less than 10 feet.

Height of a certain bridge.—The company may make the arch of the bridge for carrying the widening No.   at   over the road numbered on the deposited plans   in the parish of   of any height not less than 14 feet 6 inches.

If any bridge or viaduct of more than three arches shall intervene in any embankment, or if any tunnel shall intervene in any cutting, the length and extreme height or depth are usually marked upon the section in figures on each of the parts into which such embankment or cutting is divided by such bridge, viaduct, or tunnel, so that at each end of the viaduct or tunnel the height from the rail level to the surface shall be written thereon, also the greatest height from the rail level to the surface and the length in yards measured horizontally to accord with the plan.

Where tunnelling, as a substitute for open cutting, or a viaduct as a substitute for solid embankment is intended, the same shall be marked on the section, and, as stated

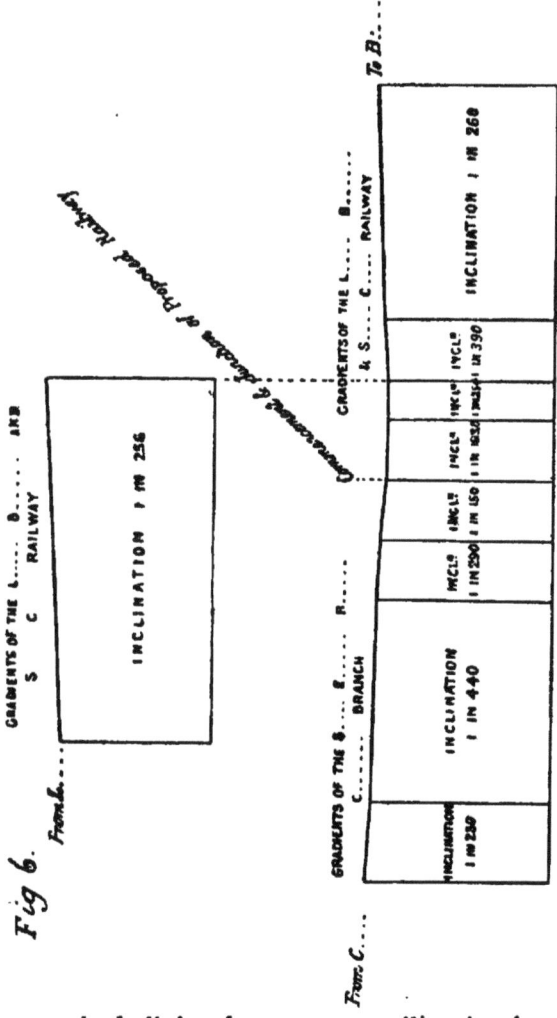

above, no work shall be shown as tunnelling in the making of which it will be necessary to cut through or remove the surface soil. The word "tunnel" should also be written along it, but covered way must not be so indicated.

Where an authorised line of railway is crossed by the proposed line, it is well to indicate the level of rails in order to show that the position of the authorised line has been duly considered, and to figure the height from the datum as authorised, taking it from the deposited section of the authorised railway. Where a tramway is crossed the surveyor should figure the height on the section in the same way.

Fig. 6 (page 274) gives an example of the half-mile sections which have to be taken, when a railway is intended to form a junction with an existing or an authorised line of railway.

In taking a longitudinal section for a railway over country of variable level, where the whole line cannot be chained from end to end, it is very important to have a correct plan upon which the centre line is drawn. The distance at which the centre line crosses each fence can be scaled from the intersection of fences, a gate, or some other defined point, and measured in the field with the use of a chain or with the employment of a tape and arrows. The centre line between two fences in which points have been thus determined can then be chained and levels taken at any required points. If the line be curved between such fences, the line joining them can be used when so marked out as a base-line for offsets scaled upon the plan, from which to set out the centre line required, and laths inserted to guide the chain. The position of points set out in the fences can be marked temporarily with pieces of newspaper, placed in the hedge branches or in the joints of a fence, by which means the formidable action of carrying ranging rods is dispensed with. Frequent temporary bench marks should be made along the section by taking the level of several fixed points as near the centre line as possible, so that in the event of any alteration in the centre line being found necessary, the fresh portion of a new section can be easily connected to the plotting of the old section. The top of the bottom hooks to gates in a fence are frequently called into requisition for temporary bench marks in this way. (See page 173.) The centre line as set out should be based

PARLIAMENTARY SURVEYING.

upon a trial section, furnishing high and low points obtained from a series of flying levels fixed in position on the plan. The trial section is usually plotted from spot levels specially arranged to suit the crossing of roads.

In cases of river bills for improving the navigation of any river, there should be a section which should specify the levels of both banks of such river; and where any alteration is intended to be made therein, this section must describe the same by feet and inches, or decimal parts of a foot.

GENERAL SUMMARY (SECTION).

The following data should be read through in the completion of every railway section, in order to ascertain that the requirements of the standing orders for deposited sections have not been overlooked in any particulars:—

1.—The datum to be described.
2.—The surface of ground, describe.
3.—Line corresponding with upper surface of rails.
4.—Commencement of railway, junction with, &c.
5.—Miles and furlongs on the datum.
6.—Termination of railway, length in miles, furlongs, and chains.
7.—Junction with railways and branches.
8.—Section of 800 yards at junctions each way.
9.—Gradients, heights at ends.
10.—Banks and cuts over 5 feet.
11.—Heights of roads above or below rails in feet and inches, calculate from gradients.
12.—Note all roads and streets as "roads."
13.—Describe (*a*) if altered in level, (*b*) if diverted, (*c*) if unaltered. Reference to cross section number.
14.—Cross sections level of road when road altered (*a*) rail level, (*b*) present inclination, (*c*) when altered. (See page 272.)
15.—Height and span of arches.
16.—Lengths of tunnels and viaducts.
17.—Span and heights of nearest bridges over canals and navigable rivers.
18.—Scales.
19.—Tidal rivers' level high and low tidal waters.

20.—Cut line.
21.—Tunnel of existing railway or canal.
22.—Level of rails in lines authorised.
23.—Gradients at junctions to correspond.

Deposits on or before the 30th November.—In cases of bills of the second class, a plan and also a duplicate thereof, together with a book of reference thereto, and a section and also a duplicate thereof, as hereinafter described, and in cases of bills of the first class by which any lands or houses are intended to be taken, a plan and duplicate thereof, together with a book of reference thereto, shall, under the Standing Orders, be deposited for public inspection at the office of the clerk of the peace for every county, riding, or division in England or Ireland, or in the office of the principal sheriff clerk of every county in Scotland, and where any county in Scotland is divided into districts or divisions, then also in the office of the principal sheriff clerk in or for each district or division in or through which the work is proposed to be made, maintained, varied, extended, or enlarged, or in which such lands or houses are situate.

In the case of railway bills, the Ordnance map, on the scale of 1 inch to a mile, or where there is no Ordnance map, a published map, to a scale of not less than half an inch to a mile (or in Ireland, to a scale of not less than a quarter of an inch to a mile), with the line of railway delineated thereon, so as to show its general course and direction, has to be deposited with such copy of plan, section, and book of reference; and in the case of railway bills a copy also has to be deposited in the Private Bill Office of the House of Commons, and in the office of the Clerk of the Parliaments (House of Lords). Further deposits with urban district councils and generally as required by the Standing Orders must receive attention.

The House of Commons provides for the clerks of the peace or sheriff clerks or their respective deputies, to make a memorial in writing upon the plans, sections, and books of reference deposited with them, denoting the time at which the same were lodged in their respective offices, and that at all reasonable hours of the day any person may examine one of the same, and make copies or extracts therefrom. *One* of the two plans and sections so deposited is sealed up and retained in the possession of the clerk of the peace or sheriff clerk until called for by order of one of the two Houses of Parliament. In cases of bills whereby it is proposed to alter or extend the municipal boundary of any city, borough, or urban sanitary district, a map on a scale of not less than 3 inches to a mile, and also a duplicate thereof, showing as well the present boundaries of the city, borough, or urban sanitary district as the boundaries of the proposed extension, is deposited with the town clerk of such city or borough, or clerk of such urban sanitary district, who shall at all reasonable hours of the day permit any person to view and examine such map, and to make copies thereof. Thus surveyors are often called upon to make copies of deposited plans and sections at these offices; or in London, at the Private Bill Office, for the use of opponents to the proposed undertaking. At the latter office a fee of five shillings is usually paid for inspection, and a fee of one sovereign for the privilege of tracing the whole or any portion of a deposited plan and section.

The Standing Orders require in cases affecting harbours, tidal waters, or navigation, where the work is to be situate on tidal lands within the ordinary spring tides, that a copy of the plans and sections shall, on or before the 30th day of November immediately preceding the application for the bill, be deposited at the office of the Harbour Department, Board of Trade, marked "Tidal Waters," and on such copy all tidal waters shall be coloured blue, and if the plans include any bridge across tidal waters the dimensions must be obtained by the surveyor, as regards span and headway of the nearest bridges, if any, across the same tidal waters above and below the proposed new bridge. Every crossing of a railway and tramway should be specified as to the crossing, over, under, or on the level, and is marked upon the deposited copy, and in all such cases such plans and sections are accompanied by an Ordnance or published map of the country over which the works are proposed to extend, or are to be carried, with their position and extent or route accurately laid down thereon.

In the case of railway, tramway, subway, and canal bills, a copy of all plans, sections, and books of reference required to be deposited in the office of any clerk of the peace or sheriff clerk (and in the case of railway bills also a copy of the said Ordnance or published map, with the line of railway delineated thereon), shall, on or before the 30th day of November immediately preceding the application for the bill, be deposited in the office of the Board of Trade.

As to conservators, and in cases where the work is to be situate on the banks, foreshore, or bed of any river, having a board of conservators constituted by Act of Parliament, a copy of the plans and sections is, on or before the 30th day of November immediately preceding the application for the bill, to be deposited at the office of the conservators of that river; and if the plans include any tunnel under, or bridge over the river, the dimensions as regards depth below bed of the river, and span and headway shall be marked thereon, and such plans are, as in the case of harbour, tidal water, or navigation schemes, accompanied by an Ordnance or published map of the country over which the works are proposed to extend, or are to be carried, with their position and extent or route accurately laid down thereon.

The span and headway are noted upon the section, and in case the section is not upon the same sheet as the plan, the surveyor writes upon the plan:—"NOTE.—For span and headway see sheet No.   of section   ." It is sufficient to name 10 feet headway for inland navigation, leaving the actual additional headway to be granted as a matter of agreement with the authorities during construction.

On or before the 30th day of November, in cases where any portion of the work shall be situate within the limits of the metropolis, as originally defined by "The Metropolis Management Act, 1855," a copy of so much of the plans and sections as relates to such portion of the work shall be deposited at the office of the London County Council, Spring Gardens.

Parish Deposits.—It is also prescribed that on or before the 30th day of November, a copy of so much of the said plans and sections as relates to each parish in or through which the work is intended to be made, maintained, varied, extended or enlarged, or in which any lands or houses, intended to be taken, are situate, together with a copy of so much of the book of reference as relates to such parish, shall be deposited with the parish clerk of each such parish in England, or, in the case of any extra-parochial place, with the parish clerk of some parish immediately adjoining thereto, or in case of any place within the limits of the metropolis, as originally defined by the "Metropolis Management Act, 1855," with the clerk of the vestry of each parish in schedule A., and with the clerk of the district board of parishes, in Schedule B. of the said Act, with the session clerk of each such parish in Scotland, and in royal burghs with the town clerk, and with the clerk of the union within which such parish is included in Ireland.

The making up of the parish deposits from the printed sheets, as far as applicable, is a responsible duty generally executed at the lithographer's office under the special direction of the engineer and the solicitor.

Deposits at the Home Office.—The following deposit of plans, &c., at the Home Office, must receive attention. Where, by any bill, power is sought to take any churchyard, burial ground, or cemetery, or any part thereof, or to disturb the bodies interred therein, or where power is sought to take any common or commonable land as the case may be, a copy of so much of the plans, sections, and books of reference required by the Standing Orders to be deposited in the Private Bill Office of the House of Commons, or in the office of the Clerk of the Parliaments (House of Lords), in respect of such bill as relates to such churchyard, burial ground, or cemetery, common or commonable land, shall, on or before the 30th day of November, be deposited at the office of the Secretary of State for the Home Department, and a copy of so much of the said plan, section, and book of reference as relates to such commonable land, shall on or before the said date be deposited at the office of the Board of Agriculture. The order in the House of Commons adds a similar provision for bills in which power is sought to take any common or commonable land as the case may be. With regard to sanitary authorities, it is prescribed that on or before the 30th day of November, a copy of so much of the said plans and sections as relates to the district of any urban sanitary authority in England or Ireland, in or through which the work is intended to be made, maintained, varied, extended, or enlarged, or in which any lands or houses intended to be taken, are situate, together with a copy of so much of the books of reference as relate to that district, shall be deposited with the clerk of that sanitary authority.

Deposit of tramway map at the office of Board of Trade.— On or before the 30th November, in the case of application for bills or provisional orders to lay down a street tramway, a published map of the district on a scale of not less than six inches to a mile (or if no map on such a scale be published, then the best map obtainable), with the line of the proposed tramway marked thereon, and a diagram on a scale of not less than two inches to a mile has to be prepared for deposit at the office of the Board of Trade.

Gas and Water.—The map required to be deposited on or before the 30th day of November by the Gas and Waterworks Facilities Act, 1870, must be on a scale of not less than six inches to the mile, and must show distinctly the situation of the land proposed to be used for the manufacture of gas, or of residual products arising in the manufacture of gas, in relation to the adjoining lands and premises, and to the urban sanitary district generally. The plan and section of any proposed new works required to be deposited on or before the day aforesaid must be on a scale of not less than one inch to the 50 feet, and must, as far as practicable, show the general arrangement, elevation, and character of the proposed works. In any case where any part of the works would be situate on lands where the ordinary spring tide flows, the site of such lands must be coloured blue on the map and plan. A copy of such map and plan coloured in like manner, and marked "Tidal Waters," shall be deposited at the Board of Trade on or before the date last aforesaid.

Under the Board of Trade regulations, the plans and sections to be deposited in all cases of application for provisional orders must be such plans and sections as are required by the Standing Orders of the House of Commons, when an Act is being applied for by means of a private bill. The Ordnance sheet (or if no Ordnance sheet is published, a general key map), showing the exact position of the proposed works, is also required to be deposited. Duplicates of the deposits made at the Board of Trade have at the same time to be deposited in the Private Bill Office of the House of Commons and in the office of the Clerk of the Parliaments, House of Lords.

Notices and applications to owners, lessees, and occupiers of lands and houses.—On or before the 15th day of December, immediately preceding the application for a bill by which any land or houses are intended to be taken, or an extension of the time granted by any former Act for that purpose is sought, application in writing has to be made to the owners or reputed owners, lessees or reputed lessees, and occupiers of all lands and houses so intended to be taken, or which may be taken as being within the limits of deviation defined upon the plan.

In cases of bills of the second class in the Commons, or of local bills of the second class in the Lords, such application must be, as nearly as may be, in the form provided by the standing orders, which notice states that application is intended to be made to Parliament in the ensuing session for a certain Act, and that the property mentioned in the schedule annexed (pages 283, 284), or some part thereof—in the case of important buildings referenced as warehouses or manufactories in towns, it is important for the solicitors to state in the notice whether "the whole" or only "a portion" referred to under a stated number is needed to be taken—in which the party addressed is believed to be interested as therein stated, will be required for the purposes of the said undertaking, according to the line thereof as at present laid out, or may be required to be taken under the usual powers of deviation to the extent of   yards on either side of the said line, which will be applied for in the said Act. Also, that a plan and section of the said undertaking, with a book of reference thereto, have been or will be deposited (as provided by the standing orders) at the place named in this notice. Further, that the party so addressed may be required to sell and convey a part only of the property numbered   on the deposited plans, and that as the promoters of the undertaking are required to report to Parliament whether assent or dissent from the proposed undertaking is expressed, or whether the party named is neuter in respect thereto, an answer of assent, dissent, or neutrality in the form left therewith is requested to be returned duly signed on or before   the   day of   next; and if there should be any error or misdescription in the annexed schedule, it is requested that information necessary to correct the same may be forwarded without delay.

SCHEDULE REFERRED TO IN THE FOREGOING NOTICE, DESCRIBING THE PROPERTY THEREIN ALLUDED TO.

	Parish, Township, Townland, or extra-parochial Place.	Number on Plans.	Description.	Owner.	Lessee.	Occupier.
Property on the Line of the proposed Work, or within the Limits of the Deviation intended to be applied for.						

The first column is now usually divided into two headings :—(1) Property to be taken compulsorily. (2) Property on which an improvement charge can be made.

Schedule referred to in the foregoing Act, describing Lands, Buildings, and Manufactories whereof portions only are required to be taken by the Company.

No. on Deposited Plans.	Parish.	Description of Property.

Relinquishment of Act.—On or before the 15th day of December immediately preceding the application for a bill, whereby the whole or any part of a work authorised by any former Act is intended to be relinquished, notice in writing of such bill has to be served upon the owners or reputed owners, lessees or reputed lessees and occupiers of the lands in which any part of the said work intended to be thereby relinquished is situate.

Under the Board of Trade rules the notices prescribed by the Standing Orders to be served on or before the 15th day of December have to be observed in all applications for Provisional Orders.

Separate lists are subsequently made of the names of such owners, lessees, and occupiers, distinguishing those who have assented, dissented, or are neuter in respect to such application, or who have returned no answer thereto; and where no written acknowledgment has been returned to an application forwarded by post, or where such application has been returned as undelivered at any time before the making-up of such lists, the direction of the letters in which the same was so forwarded is inserted therein.

Water.—And on or before the 15th day of December, notice in writing shall be given to the owners or reputed owners, lessees or reputed lessees, and occupiers of all mills and manufactories, or other works using the water of such stream for a distance of 20 miles below the point at which such water is intended to be abstracted, such distance to be measured along the course of such stream, unless such waters shall within a less distance than 20 miles fall into or unite with any navigable stream, and then only to the owners or reputed owners, lessees or reputed lessees, and occupiers of such mills or manufactories or other works, as aforesaid, which shall be situate between the point at which such water is proposed to be abstracted and the point at which such water shall fall into or unite with such navigable stream. Hence it is necessary for the surveyor to examine the course of such streams. The notice shall state the name (if any) by which the stream is known at the point at which such water shall be immediately abstracted, and also the parish in which such point is situate, and the time and place of deposit with the clerks of the peace and sheriff clerks, as the case may be, for reference to the plans, sections, and books of reference, and copies of the gazette notice respectively.

Cemetery, Gas Works, Sewage Works.—On or before the 15th day of December immediately preceding the application for a bill for making a cemetery or burial ground, or for constructing gasworks or sewage works, or works for the manufacture or conversion of the residual products, notice shall be served upon the owner, lessee, and occupier of every dwelling-house situate within *three hundred yards* of the limits within which such cemetery or burial ground may be made or such works may be constructed, and the surveyor is called upon to report the position of such houses so situated. In the case of a provisional order similar notice must be given when works are proposed to be constructed for the storage of gas.

Under the Board of Trade rules provision is made as in the case of tramways for notice to be given to owners and lessees of railways, tramways, and canals on or before the 15th day of December immediately preceding the application for any provisional order for constructing gasworks or waterworks, or works connected therewith whereby it is proposed to lay any pipes along or across any railway, tramway, or canal, or otherwise to affect or interfere with such railway, tramway, or canal, notice in writing of such application shall be served upon the owner or reputed owner, and upon the lessee or reputed lessee of such railway, tramway, or canal, and such notice shall state the place or places at which the plan of the works to be authorised by such order has been or will be deposited. Every such notice must state that any clause which such owners or lessees may desire to have inserted in the order must be submitted to the Board of Trade on or before the 15th February.

Tramway.—On or before the 15th day of December immediately preceding the application for a bill or provisional order for laying down a tramway, notice in writing has also to be given to the owners or reputed owners, lessees or reputed lessees, and occupiers of all houses, shops, or warehouses abutting upon any part of the said highway, where, for a distance of 30 feet or upwards, a less space than that specified in page 264 shall intervene between the outside of the footpath on either side of the road and the nearest rail of the tramway.

On or before the 15th day of December immediately preceding the application for any bill or provisional order for laying down a tramway crossing any railway or tramway on the level, or crossing any railway, tramway, or canal by means of a bridge, or otherwise affecting or interfering with such railway, tramway, or canal, notice in writing of such application is served upon the owner or reputed owner, and upon the lessee or reputed lessee of such railway, tramway, or canal, and such notice shall state the place or places at which the plans of the tramway to be authorised by such bill have been, or will be, as required, deposited.

The Board of Trade rules provide that in the case of a provisional order, any such notice must state that any clause which such owner or lessee may desire to be inserted in the order must be submitted to the Board of Trade on or before the 15th February.

Bill.—On or before the 17th December, a printed copy of every local bill, proposed to be introduced into either House of Parliament, is deposited by the Parliamentary agent in the office of the Clerk of the Parliaments; and on or before the 21st December, every petition for a private bill, headed by a short title descriptive of the undertaking, corresponding with that at the head of the advertisement, together with a declaration signed by the agent, and a printed copy of the bill annexed, is deposited in the Private Bill Office of the House of Commons; and such petition, bill, and declaration shall be open to the inspection of all parties; and printed copies of the bill are also delivered therewith for the use of any member of the House or agent who may apply for the same. In such declaration the agent states to which of the two classes of bills such bill, in the judgment of the agent, belongs.

And if the proposed bill shall give power to effect any of the following objects, that is to say :—(1) Power to take any lands or houses compulsorily, or to extend the time granted by any former act for that purpose. (2) Power to levy tolls, rates, or duties; or to alter any existing tolls, rates, or duties; or to confer, vary, or extinguish any exemption from payment of tolls, rates, or duties, or to confer, vary, or extinguish any other right or privilege. (3) Power to amalgamate with any other company, or to sell or lease their undertaking, or to purchase or take on lease the undertaking of any other company. (4) Power to interfere with any crown, church, or corporation property or property held in trust for public or charitable purposes. (5) Power to relinquish any part of a work authorised by a former Act.

(6) Power to divert into any existing or intended cut, canal, reservoir, aqueduct, or navigation, or into any intended variation, extension, or enlargement thereof respectively, any water from any existing cut, canal, reservoir, aqueduct, or navigation whether directly or derivatively, and whether under any agreement with the proprietors thereof or otherwise. (7) Power to make, vary, extend, or enlarge any cut, canal, reservoir, aqueduct, or navigation. (8) Power to make, vary, extend, or enlarge any railway. Then the said declaration shall state which of such powers are given by the bill, and shall indicate in which clauses of the bill (referring to them by their number) such powers are given, and shall further state that the bill does not give power to effect any of the objects enumerated in this order other than those stated in the declaration.

If the proposed bill shall not give power to effect any of the objects enumerated in the preceding order, the said declaration shall state that the bill does not give power to effect any of such objects. The said declaration shall also state that the bill does not give any powers, other than those included in the notices for the bill.

The clerks in the Private Bill Office of the House of Commons are particularly directed to take care, that in the examination of all private bills levying any rates, tolls, or duties on the subject, the names of Peers of Parliament, Peers of Scotland, or Peers of Ireland, are not to be inserted therein, either as trustees, commissioners, or directors of any company, except where such rates, tolls, or duties are made or imposed for services performed, and are not in the nature of a tax.

On or before the 21st day of December, a printed copy of every private bill (Commons), or of every local bill (Lords), shall be deposited at the office of Her Majesty's Treasury, and at the General Post Office; in the case of every railway, tramway, subway, gas, water, patents, electric lighting, and canal bill, and of every bill for incorporating or giving powers to any company, shall be deposited at the office of the Board of Trade. Also for every bill relating to any dock, harbour, navigation, pier, or port, or tidal water a copy shall be deposited at the office of the Fisheries and

Harbour Department of the Board of Trade, marked "Tidal Waters." Also for every bill whereby power is sought to take any churchyard, burial ground, or cemetery, or any part thereof, or to disturb the bodies interred therein, shall be deposited at the office of the Secretary of State for the Home Department. Also for every bill whereby application is made by, or on behalf of, any municipal corporation or local authority in England or Wales, for power in respect of any purpose to which the several acts specified in part 1 of the schedule to "The Local Government Board Act, 1871," relate, and of every bill relating to turnpike roads or trusts, highways or bridges, shall be deposited at the office of the Local Government Board. Also in the case of every bill of the second class, whereby any work shall be authorised within the limits of the metropolis, as originally defined by "The Metropolis Management Act, 1855," shall be deposited at the office of the London County Council. And in the case of every bill of the second class, whereby it is intended to authorise the construction of any work on the banks, foreshore, or bed of any river having a board of conservators constituted by Act of Parliament, shall be deposited at the office of the Conservators of the river.

On the 22nd day of December, between the hours of twelve and three, Parliamentary agents are allowed to exchange by agreement the numbers originally assigned to their petitions, after which date, the final General List of Petitions is made out in the Private Bill Office, in which the petitions are numbered consecutively from one to the highest number, according to the order in which they have been finally entered in the register. This list is subsequently printed, and a copy can be obtained upon application at the Private Bill Office.

In the case of provisional orders, the promoters are to deposit at the office of the Board of Trade on or before the 23rd December :—

1. "A memorial of the promoters, signed by them or one of them, headed with a short title descriptive of the undertaking or application (corresponding with that at the head of the advertisement), addressed to the Board of Trade, and praying for a provisional order.

2. "A printed draft of the provisional order as proposed by the promoters printed upon one side of the page of the paper so as to leave the back of the page blank, and care must be taken that any schedule annexed thereto begins a new page.

3. "An estimate of the expense of the proposed new works, if any, signed by the person making the same.

"They are also to deposit printed copies of each draft provisional order for public inspection at the Custom House (if any) of the port, sub-port, or creek to which the application relates, and also as in the case of Light Railways, a sufficient number of such printed copies at the office named in that behalf in the advertisement; such copies to be there furnished to all persons applying for them at the price of not more than one shilling each." (See page 267.)

Under the Board of Trade regulations the following documents, &c., must be deposited in the case of application for a tramway, on the 23rd December, at the Board of Trade :—

(1.) A complete list of every railway, tramway, and canal proposed to be crossed or otherwise affected or interfered with, together with the names and addresses of the owners or reputed owners and lessees or reputed lessees thereof, and certified copy of the notice served upon them.

(2.) A complete list of the local authorities through whose districts the proposed tramway is to pass (including in such list the clerk to the Justices of the Peace in cases where it is proposed to cross county bridges), and if any such district is or forms part of a highway district, under the provisions of "The Highway Acts," a statement to that effect must accompany the deposit. (Where there is a road authority distinct from the local authority the name of such road authority must also be given.)

Fee. — Under the Board of Trade regulations the promoters of a provisional order are further required to pay a fee of £35 towards the expense of settling the provisional order; this sum must be paid to the officer of the Board of Trade with whom the draft provisional order is deposited, and at the time of the deposit.

The draft provisional order which is to be deposited at the Board of Trade on or before the 23rd of December must be deposited in triplicate, accompanied by a copy of every private or local act, charter, deed, or other document of a like nature which is referred to therein, or which will be affected thereby.

Estimate of Expense.—Engineers' estimates and promoters' declarations, with such lists of owners, lessees, and occupiers, which are required by the Standing Orders of the House, shall be deposited in the Private Bill Office of the House of Commons, and in the office of the Clerk of the Parliaments (House of Lords) on or before the 31st day of December.

On or before 31st December copies of the estimate of expense of the undertaking proposed by the bill, and where a declaration alone, or declaration and estimate of the probable amount of rates and duties, are required, copies of such declaration, or of such declaration and estimate, shall be printed at the expense of the promoters of the bill, and delivered at the Vote office for the use of the members of the House of Commons, at the Private Bill Office (House of Commons) for the use of any agent who may apply for the same, and at the office of the Clerk of the Parliaments for the use of the House of Lords.

The estimate for any works proposed to be authorised by any railway, tramway, subway, canal, dock or harbour bill, is ordered to be in the following form, or as near thereto as circumstances may permit :—

Session.........
Estimate of the proposed .... (Railway)
....

Line No. .	Miles. f. ch.	Whether Single or Double.
Length of Line		

	Cubic yds.	Price per yd.	£ s. d.	£ s. d.
Earthworks:				
Cuttings—Rock	—	—		
Soft Soil				
Roads				
Total				

Embankments, including Roads   Cubic yds.
Bridges—Public Roads............... Number
  Accommodation Bridges and Works ....
Viaducts..
Tunnels ....
Culverts and Drains ....
Metallings of Roads and Level Crossings ....
Gatekeepers' Houses at Level Crossings ....
Permanent Way, including Fencing:

Miles. f. ch.	Cost per Mile. £ s. d.
at	

Permanent Way for Sidings, and Cost of Junctions
Stations ....

Contingencies............................per cent.
Land and Buildings
    A.   R.   P.

TOTAL......£

The same details are employed for each branch and for the general summary of total cost. The first great item of expense in the construction of a Railway is the land, which is generally valued as accommodation land, consequently it is reckoned at a higher price per acre than would be applicable to a purchase of extent. The consideration of severance is always admitted in the value, notwithstanding that provision is made to compel a Company to take more land than required, when reduced to a certain small quantity.

In the case of any bill by which power is sought to take by compulsion, or by agreement, twenty or more houses in any metropolitan parish, or ten or more houses inhabited by labouring classes outside the metropolis, the promoters are ordered to deposit in the Private Bill Office (House of Commons) and in the office of the Clerk of the Parliaments (House of Lords), on or before the 31st day of December, a statement of the number, description, and situation of such houses, the number (so far as then can be ascertained) of persons to be displaced, and provision is made in the bill for remedying any inconvenience likely to arise from such displacement, and such statement shall stand referred to the Committee on the bill. A copy of so much of the plan (if any) as relates thereto has also to be prepared by the surveyor who values the Land and Buildings to be deposited at the same time.

TRAMWAYS.

Under the Board of Trade rules for provisional orders notice to all such frontagers has to be given, and the notice must contain a notification that dissent from the tramway being so laid may be stated in writing addressed to the assistant secretary of the railway department of the Board of Trade, on or before the 1st of January next ensuing, and that he must at the same time send a copy of his dissent to the promoters.

In the case of Provisional Orders, proofs of compliance with the Tramways Act, 1870, and rules of the Board of Trade are made by the Agents by the 1st January. Six days' notice will be given of the day and hour at which the

Agents are to attend for the purpose at the Board of Trade, and printed forms of proofs will accompany the notice. One of these forms should be filled up by the Parliamentary Agents and brought, with the requisite documents, to the Department at the time fixed for proof.

Standing Order Examiners.—One or more officers in each House of Parliament act to see that the Standing Orders have been complied with. They are called in the Lords "the examiners of standing orders for private bills," appointed by the House; and called in the Commons "the examiners of petitions for private bills," appointed by Mr. Speaker. The promoters of each bill are required to prove compliance with the Standing Orders of both Houses at the time appointed by these examiners. At the Private Bill Office of the House a register is kept, with blank lines numbered consecutively from 1 to 500, in which the Parliamentary agent enters the name of the petitions lodged by him, on such of the lines, not then having any petition entered thereon, as he thinks fit; and if he do not prescribe any order for the entry of such petitions, they are entered in the order in which they are deposited, upon the earliest consecutive lines then remaining unoccupied, and confirmed at the Parliamentary Agents' Conference held upon December 22nd, as previously stated.

Not less than seven clear days' notice is given in the Private Bill Office of the day appointed for the examination of each petition, and the day so appointed is written against the several petitions upon a special copy of the printed copy of the "general list of petitions," which is kept in the Private Bill Office for that purpose.

The petitions are set down for hearing before the examiners in the order in which they stand in the "general list of petitions," precedence being given whenever it may be necessary to unopposed petitions.

The examination of the petitions for bills which shall have been duly deposited in the Private Bill Office (House of Commons), and in the office of the Clerk of the Parliaments (House of Lords), commences on the 18th day of January. In case the promoters do not appear at the time when the petition comes on to be heard, the examiner to whom the case is allotted strikes the petition off the "general list of petitions," and the same cannot then be re-inserted except by order of the House.

The examiner shall certify by indorsement on each petition whether the Standing Orders have or have not been complied with ; and, when they have not been complied with, he shall also report to the House the facts upon which his decision is founded, and any special circumstances connected with the case.

Memorials complaining of non-compliance with the Standing Orders (of either House), applicable previously to the introduction of private bills, must be deposited in the Private Bill Office, House of Commons, as follows :—

If the same relate to bills numbered in the general list published by the Private Bill Office of the House of Commons as follows :—

From 1 to 100	They must be deposited by the Parliamentary agent before two o'clock on	Jan. 9.
,, 101 to 200		,, 16.
,, 201 and upwards		,, 23.

And in the House of Lords every memorial complaining of non-compliance with the Standing Orders in respect of any bill referred to the examiners after first reading, or in respect of any petition for additional provision, must, together with two copies thereof, be deposited in the office of the Clerk of the Parliaments before twelve o'clock on the day preceding that appointed for the examination.

The examiner may admit affidavits in proof of the compliance with the Standing Orders or may require further evidence.

The printed statements of proofs can be obtained at the Queen's printers, and forms a portion of the Parliamentary agent's work ; and where lists are annexed to affidavits, his name is to be entered in the statement of proofs as delivering in such lists, followed by the names of the witnesses proving the service of notices or deposit of documents, as the case may be.

With regard to memorials complaining of non-compliance with Standing Orders, the Standing Orders of both Houses provide that any parties shall be entitled to appear and to be heard, by themselves, their agents and witnesses, upon a memorial addressed to the Examiner, complaining of a non-compliance with the Standing Orders, provided the matter complained of be specifically stated in such memorial, and the party (if any) who may be specially affected by the non-compliance with the Standing Orders have signed such memorial, and shall not have withdrawn his signature thereto.

The promoters of a Provisional Order must be prepared to prove compliance by the 10th January with the provisions of the Act concerning the Standing Order requirement for deposit and publication. Six days' notice will be given of the day and hour at which the promoters are to attend for the purpose at the Board of Trade. Printed forms of proof will be supplied, which should be filled up by the promoters, and brought with the requisite documents to the Board of Trade at the time fixed for proof.

Any objection to the provisional order which it is intended to urge on the Board of Trade must be sent in to that office before the 15th January. A copy of such objections must also be sent at the same time to the promoters; and in forwarding the objections to the Board of Trade, the objectors or their agents should state that this has been done. It is also established for the confirmation of the order by Act of Parliament that an order of the Board of Trade under this part of this Act shall not of itself have any operation, but the same shall have full operation, when, and as confirmed by Act of Parliament, with such modifications, if any, as to Parliament may seem fit.

In the House of Lords at the commencement of every Session, a Standing Order Committee is appointed, consisting of forty lords, besides the Chairman of Committees, who shall be always Chairman of such Standing Order Committee. Three of the lords so appointed, including the Chairman, form a quorum in all opposed cases.

Three clear days' notice is given of the meeting of the Standing Order Committee.

All certificates from the examiners in respect of bills

in which they shall certify that the Standing Orders have not been complied with are referred to the Standing Order Committee, and the Committee report to the House whether the Standing Orders ought or ought not to be dispensed with, and in the former case, upon what terms and conditions, if any. All special reports from the examiners as to the construction of a Standing Order are referred to the Standing Order Committee, and the Committee determine, according to their construction of the Standing Order, and on the facts stated in the report, whether the Standing Orders have, or have not been complied with, and they report accordingly to the House, and if the Committee report that a Standing Order has not been complied with, they also report whether such Order ought to be dispensed with, and upon what terms and conditions, if any. When an Examiner's certificate or special report is referred to the Standing Order Committee, the Committee, if they think fit, hear the parties affected by any Standing Order referred to in such certificate or special report, provided such parties have deposited in the office of the Clerk of the Parliaments, not later than three o'clock on the day before the day on which the Committee is appointed to meet, a statement (to be printed in all opposed cases) of the facts to be submitted to the Committee. Such statement shall be confined strictly to the points reported upon by the Examiner, and no party on the consideration thereof by the Committee shall be allowed to travel into any matter not referred to in his statement.

The Examiner in the House of Commons in due course makes a report of the several cases in which he shall have certified that the Standing Orders have or have not been complied with in respect of any bills which, in pursuance of any report from the Chairman of the Committee of Ways and Means, shall originate in the House of Lords; and where they have not been complied with, he also reports, separately, the facts upon which his decision is founded, and any special circumstances connected with the case.

In case the Examiner feels doubts as to the due construction of any Standing Order in its application to a particular case, he makes a special report of the facts without deciding whether the Standing Order has or has not been complied with; and in such case he endorses the petition with the words "Special Report," either alone, or if non-compliances with other Standing Orders shall have been proved, he adds, the words "Standing Orders not complied with."

The Chairman of Ways and Means in the House of Commons, with not less than three other persons appointed by Mr. Speaker for such period as he shall think fit, are constituted Referees of the House on private bills; such Referees forming one or more courts, three at least to be required to constitute each court, provided that the Chairman of any second court shall be a member of the House of Commons, and provided that no such Referee, if he be a member of the House of Commons, shall receive any salary.

The Referees decide upon all petitions against private bills, or against Provisional Orders or provisional certificates, as to the rights of petitioners to be heard upon such petitions, without prejudice, however, to the power of the Select Committee to which the bill is referred to decide upon any question as to such rights arising incidentally in the course of their proceedings. Only the costs of one counsel are allowed for appearance before such Referees in support of a private bill, or in support of any petition in opposition thereto, unless specially authorised by the Referees.

The Select Committee to which any bill has been referred may, subject to the approval of the Chairman of Ways and Means, refer any question arising in the course of their inquiry, which they may deem suitable to be so referred, to the Referees for their decision, such question being stated in writing, and signed by the Chairman of the Committee. The Referees, so soon as their inquiry has been completed, return the question, with their decision certified thereon, to the Chairman.

The Proceedings of the Select Committee on Standing Orders are conducted by a Committee in the Commons consisting of eleven members, nominated at the commencement of every session, of whom five shall be a quorum.

When any report of the Examiner of Petitions for private bills, in which he reports that the Standing Orders have not been complied with, has been referred to the Select Committee on Standing Orders, and after the petition for the bill has been duly presented, they report to the House whether such Standing Orders ought or ought not to be dispensed with, and whether in their opinion the parties should be permitted to proceed with their bill, or any portion thereof, and under what (if any) conditions.

The Select Committee on Standing Orders have the power to report on the cases referred to them in respect of private bills originating in the House of Lords, notwithstanding that the petitions for the same have not been presented to the House.

When any special report from the Examiner of Petitions as to the construction of a Standing Order has been referred to the Select Committee on Standing Orders, they then determine, according to their construction of the Standing Order, and on the facts stated in such report, whether the Standing Orders have or have not been complied with, and they afterwards either report to the House that the Standing Orders have been complied with, or shall proceed to consider the question of dispensing with the Standing Orders, as the case may be.

When any petition praying that any of the sessional or Standing Orders of the House relating to private bills may be dispensed with, stands referred to the Select Committee on Standing Orders, they shall report to the House whether such sessional or Standing Orders ought or ought not to be dispensed with.

When any petition for the re-insertion of any petition for a private bill in the general list of petitions stands referred to the Select Committee on Standing Orders, they report to the House whether in their opinion such petition ought or ought not to be re-inserted, and, if re-inserted, under what (if any) conditions.

Estimates and Deposit of Money, and Declarations in Certain Cases previous to the 15th January.—The estimate of the expense of the undertaking under each bill of the second class being made and signed by the person making the same, a deposit has to be made prior to the 15th day of January, in the case of a railway bill, tramway bill, or subway bill authorising the construction of works by other than an existing railway company, tramway company, or subway company incorporated by Act of Parliament, possessed of a railway, tramway, or subway already opened for public traffic, and which has during the year last past paid dividends on its ordinary share capital, and which does not propose to raise under the bill a capital greater than its existing authorised capital, a sum not less than five per cent. on the amount of the estimate of expense, or in the case of substituted works, on the amount by which the expense thereof will exceed the expense of the works to be abandoned, and in the case of all bills other than railway bills, tramway bills, and subway bills, a sum not less than four per cent. on the amount of such estimate, or of such excess as aforesaid, shall, previously to the 15th day of January, be deposited with the Paymaster-General for and on behalf of the Supreme Court of Judicature in England, if the work is intended to be done in England; or with the Paymaster-General for and on behalf of the Supreme Court of Judicature in England or with the Queen's and Lord Treasurer's Remembrancer on behalf of the Court of Exchequer in Scotland, if the work is intended to be done in Scotland; or with the Accountant-General of the Supreme Court of Judicature in Ireland, if the work is intended to be done in Ireland.

Cases wherein a declaration is deposited, in which no deposit is required, occur where the work is to be made, wholly or in part, by means of funds, or out of money to be raised upon the credit of present surplus revenue, belonging to any society or company, or under the control of directors, trustees, or commissioners, as the case may be, of any existing public work, such parties being the promoters of the bill, a declaration stating those facts, and setting forth the nature of such control, and the nature and amount of such funds or surplus revenue, and showing the actual surplus of such funds or revenue, after deducting the funds required for purposes authorised by any Act or Acts of Parliament, and also the funds which may be required for any other work to be executed under any bill in the same session, and given under the common seal of the society or company, or under the hands of some authorised officer, of such directors, trustees or commissioners, may be deposited, and in such case no deposit of money is required in respect of so much of the estimate of expense as shall be provided for by such surplus funds.

Cases wherein a declaration and an estimate of rates may be deposited will transpire in cases of any bill under which no private or personal pecuniary profit or advantage is to be derived, and where the work is to be made out of money to be raised upon the security of the rates, duties or revenue already belonging to or under the control of the promoters, or to be created by or to arise under the bill, a declaration stating those facts, and setting forth the means by which funds are to be obtained for executing the work, and signed by the party or agent soliciting the bill, together with an estimate of the probable amount of such rates, duties or revenue, signed by the person making the same, may be deposited, and in such case no money deposit is required.

In making the preliminary arrangements, the surveyor should also bear in mind the following restrictions as to mortgage. In the case of a railway bill, no company shall be authorised to raise by loan or mortgage a larger sum than one-third of their capital; and, until fifty per cent. on the whole of the capital shall have been paid up, it shall not be in the power of the company to raise any money by loan or mortgage. The House of Commons provide that the Committee upon the bill may report that such restrictions, or either of them, ought not to be enforced, with the reasons on which their opinion is founded.

The same rule shall apply in the case of a tramway bill or subway bill, one-fourth of the capital being substituted for one-third.

Proceedings of Committees on Opposed Bills.—In the House of Lords all opposed bills are placed in separate groups denoted by letters of the alphabet, to be referred to a separate Select Committee of five lords who must have no individual interest in the undertaking. Every such Committee sits at 11 a.m. of the day appointed until

4 p.m. Any alteration of the hours needs the special leave of the House. The Committee is appointed by a committee of selection consisting of the Chairman of Committees and four other lords named by the House.

In the House of Lords it is ordered that no petition praying to be heard upon the merits against any Local Bill or Provisional Order Confirmation Bill shall be received, unless the same is presented by being deposited in the Private Bill Office before 3 o'clock in the afternoon, on or before the seventh day after the day on which such Bill has been read a second time.

No petition praying to be heard upon the merits against any Local Bill or any Provisional Order Confirmation Bill brought from the House of Commons shall be received, unless the same be presented by being deposited in the Private Bill Office before 3 o'clock in the afternoon, on or before the seventh day after the day on which such bill has been read a first time.

In the House of Commons the Committee on every opposed railway and canal bill, or group of railway and canal bills, is usually composed of four members and a Referee, or four members not locally or otherwise interested in the bill or bills referred to them; the Chairman to be appointed by the General Committee on railway and canal bills, and three other members by the committee of selection. The committee so appointed generally assemble at 11.30 or 12.0 o'clock. The time is announced in the printed notices of the House.

The Chairman of the Committee of Ways and Means (House of Commons) shall, at the commencement of each Session, seek a conference with the Chairman of Committees of the House of Lords, for the purpose of determining in which House of Parliament the respective private bills should be first considered, and such determination shall be reported to the House. The Parliamentary Agents learn the decision and acquaint the promoters or their solicitors.

"The Committee of Selection" consists of the Chairman of the Select Committee on Standing Orders, who shall be *ex-officio* Chairman thereof, and seven other members, who are nominated at the commencement of every session, of which committee, three form a quorum.

"The General Committee on Railway and Canal Bills" is nominated at the commencement of every session by the committee of selection, of which committee three form a quorum.

The committee on every opposed private bill (not being a railway or canal bill), or group of bills, the groups being expressed by numbers, and the committee on any bill to confirm any Provisional Order or provisional certificate is composed of a chairman and three members and a Referee, or a chairman and three members not locally or otherwise interested in the bill or bills referred to them, to be appointed by the committee of selection.

Committees are not allowed to proceed if more than one of their members be absent, unless by special leave of the House.

When the bill has reached the Committee stage to be considered upon its merits, if it is opposed, a diagram is made or map prepared by the promoters showing the scheme to hang up in the Committee room. In the House of Commons, Committee room No. 19 measures 20 ft. 8 in. from ceiling to floor, and 14 ft. 6 in. from ceiling to dado. The width of wall between windows and doors is 28 ft. 6 in., and the width of wall along windows or doors is 33 ft. 3 in., but in Committee rooms Nos. 7 and 14, the latter width between partition rooms is 22 ft. 9 in. The width of wall between doors is 18 ft. 6 in., and this is the wall best adapted for a map as it faces the light and can be inspected without turning round, both by Members of the Committee and by the contending parties. Hence 18 ft. length by 14 ft. high or 27 ft. 6 in. maximum length by 14 ft. high are suitable dimensions for Committee diagrams, but 18 ft. × 14 ft. or less is the most convenient size.

No petition against a private bill or a bill to confirm any provisional order or provisional certificate is considered by the committee on such bill which shall not distinctly specify the ground on which the petitioners object to any of the provisions thereof; and the petitioners are only heard on such grounds so stated ; and if it shall appear to the said committee that such grounds are not specified with sufficient accuracy the committee may direct that there be given in to the committee a more specific statement in writing, but limited to such grounds of objection so inaccurately specified.

The petition must have been prepared and signed in strict conformity with the rules and orders of Parliament, and have been presented in due form and deposited in the Private Bill Office, in the case of private bills, not later than ten clear days after the first reading of such bill, and in the case of bills to confirm any Provisional Order or provisional certificate, not later than seven clear days after the Examiner shall have given notice of the day on which the bill will be examined, except where the petitioners shall complain of any matter which may have arisen during the progress of the bill before the said committee, or of any proposed additional provision, or of the amendments as proposed in the filled-up bill deposited in the Private Bill Office.

It shall be competent to the Referees on private bills to admit petitioners to be heard upon their petitions against a private bill, on the ground of competition, if they shall think fit.

No petition for additional provision can be presented to the House of Lords without the sanction of the Chairman of Committees, but no petition for additional provision shall be received in the case of a bill brought from the House of Commons.

Where a railway bill contains provisions for taking or using any part of the lands, railway, stations, or accommodations of another company, or for running engines or carriages upon or across the same, or for granting other facilities, such company are entitled to be heard upon their petition against such provisions or against the preamble and clauses of such bill.

It shall be competent to the Referees on private bills to admit the petitioners, being the municipal or other authority having the local management of the metropolis, or of any town, or the inhabitants of any town or district alleged to be injuriously affected by a bill, to be heard against such bill, if they shall think fit, but no Committee of either House shall have power to examine into the compliance or non-compliance with such Standing Orders as are directed to be proved before the Examiner, unless by special order of the House.

The municipal or other local authority of any town or district alleging in their petition that such town or district may be injuriously affected by the provisions of any bill relating to the lighting or water supply thereof, or the raising of capital for any such purpose, has a *locus standi* to be heard against such bill.

The owner, lessee or occupier of any house, shop, or warehouse in any street through which it is proposed to construct any tramway, and who alleges in any petition against a private bill or Provisional Order, that the construction or use of the tramway proposed to be authorised thereby will injuriously affect him in the use or enjoyment of his premises, or in the conduct of his trade or business, is entitled to be heard on such allegations before any Select Committee to which such private bill or the bill relating to such Provisional Order is referred.

In all cases of opposed private bills, in which no parties shall have appeared on the petitions against such bills, or having appeared shall have withdrawn their opposition before the evidence of the promoters shall have been commenced, the committees on such bills may refer them back, with a statement of the facts, if not railway or canal bills, to the committee of selection, and, if railway and canal bills, to the general committee on railway and canal bills, who shall deal with them as unopposed bills.

Every committee on a railway bill shall report specially to the House,—

Whether it be intended that the railway shall cross on a level any railway, turnpike road or highway.

And any other circumstances which, in the opinion of the Committee, it is desirable that the House should be informed of.

No railway whereon carriages are propelled by steam or by atmospheric agency, or drawn by ropes in connection with a stationary steam-engine, is allowed to be made across any railway, tramway, turnpike-road or other public carriage-way on the level, unless a report thereupon from some officer of the Board of Trade shall be laid before the Committee on the bill, and unless the committee, after considering such report, if they shall disagree with the said report, recommend such level crossing, giving the reasons and facts upon which their opinion is founded; and in every clause authorising a level crossing, the number of lines of rails authorised to be made at such crossing shall be specified.

A clause is inserted in Railway, Tramway, and Subway Bills imposing penalty unless the line be opened. This applies to every bill whereby the construction of any new line of railway, tramway, or subway is authorised, or the time for completing any line already authorised is extended.

Also for railway, tramway, or subway deposits a clause is usually inserted providing that deposits be impounded as security for the completion of the line, and a clause providing for (1) application of depositor, (2) penalty in compensation to injured parties, and (3) the time limited for completion of line.

The period limited shall not, in the case of a new railway line, exceed five years [or in the case of a new tramway line two years], and the extension of time for completion shall not in the case of a railway line exceed three years [or in the case of a tramway line one year]. In the case of extension of time the additional period shall be computed from the expiration of the period sought to be extended.

In any Railway Bill or Tramway Bill to which the preceding provisions are not applicable, the Committee on the Bill shall make such other provision as they shall deem necessary for ensuring the completion of the line of railway or tramway.

Notices to be given and deposits made in cases where work is altered while bill is in Parliament.

Whenever during the progress through either House of any bill of the second class originating in that House, any alteration has been made in any work authorised by such bill, proof must be given before the Examiners that a plan and section of such alteration, on the same scale and containing the same particulars as the original plan and section, together with a Book of Reference thereto, has been deposited in the Private Bill Office (in the case of the Commons), or in the office of the Clerk of the Parliaments (House of Lords), so that a record is kept in both Houses, also with the Clerk of the Peace of every county, riding, or division in England or Ireland, and in the office of the sheriff clerk of every county in Scotland in which such alteration is proposed to be made, and where any county in Scotland is divided into districts or divisions, then also in the office of the principal sheriff clerk in and for each district or division in which such alteration is proposed to be made; also that a copy of such plan and section, so far as relates to each parish, together with a Book of Reference thereto, has been deposited with the parish clerks of each such parish in England, or, in the case of any extra-parochial place, with the parish clerk of some parish immediately adjoining thereto, with the session clerk of each such parish in Scotland, and in royal burghs with the town clerk, and the clerk of the union within which such parish in Ireland is included, in which such alteration is intended to be made, two weeks previously to the introduction of the bill into the House; and that the intention to make such alteration has been published previously to the introduction of the bill into the House once in the London, Edinburgh or Dublin Gazette, as the case may be, and for two successive weeks in some one and the same newspaper of the county in which such alteration is situate, or if there be no such printed paper published therein, then in the newspaper of some county adjoining thereto; and that application in writing as nearly as may be in the form set forth in the Appendix marked (A) was duly made to the owners or reputed owners, lessees or reputed lessees, or in their absence from the United Kingdom, to their agents respectively, and to the occupiers of lands through which any such alteration is intended to be made; and the consent of such owners or reputed owners, lessees or reputed lessees, and occupiers, to the making of such alteration, shall be proved before the Examiner.

In conclusion, the Parliamentary surveyor must remember that the Standing Orders of both Houses of Parliament, as well as the Board of Trade rules and regulations for Provisional Orders, are subject to amendment from time to time, and that it is his duty to acquaint himself, prior to the execution of sessional work, with every alteration that has been made by the authorities and published. Any Parliamentary agent with whom he may be acquainted will gladly inform him what recent amendments may have been made since the above notes were written or during the Session of Parliament immediately preceding that for which his work is about to be prepared.

CHAPTER XXII.

RAILWAY WORK.

In railway work, prior to construction, the ground to be surveyed being long and narrow, the ordinary method of triangulation can seldom be resorted to, and a series of continuous base lines form the chief feature. Having ranged out the base lines and determined the main station points, the enclosures upon each side through which the intended railroad is to pass are surveyed by trapezium lines tied on to the base line at fixed points by diagonal lines, except in such cases where the direction of any boundary is such as to invite the formation of a triangle, one side of which lies upon the base line. Sometimes the position of the trapezium is fixed by taking the angle between the sides with a box sextant, and sometimes the trapezium lines enclosing the base line are measured, and the points of intersection with the base lines carefully noted. In open country the proposed centre line of the railway may be used as a base line, and the angles at the intersection of such base lines recorded so as to fix their position. The Parliamentary plan and section may, under certain circumstances, be enlarged and made use of for purposes of a preliminary tender.

Owing to the facilities for vision, the inexperienced surveyor in running a trial line for a railway may be tempted to place the intersection points where the curves would turn upon the several ridges crossed; but such a course of action should be studiously avoided where possible, because if such a line be adopted, the cuttings through the hills would all be curved in plan, whereas the best construction is to confine curves as much as possible to comparatively level and open ground. The diagram (pp. 313, 314) illustrates a working contract section for a line of railway, 6 furlongs 1 chain in length. The surface levels of the natural ground having been plotted by the method explained in Chapter XVI., the engineer then proceeds to rule up vertical red or blue lines upon the section at each chain's length perpendicular to the datum line. The level of the datum line having been determined, the heights of the surface levels above the datum line are then written, as shown upon these vertical lines, just above the datum line. These heights should be derived as much as possible from the figures contained in the column headed "Reduced Levels," in the "Level Book," and should be only scaled as a check upon the plotting, or at points where no reduced levels are given in the "Level Book." The level of the upper surface of the rails should next be figured upon the section. This is arrived at by first "grading" the section, by stretching a suitable length of cotton over portions of the irregular line which indicates the present level of the ground, and adapting the cotton line by successive rising or falling gradients to allow of the amounts of banks and cuttings in each case appearing equal to the eye. In so grading a section, regard must be had to what is termed the "ruling gradient," namely, 1 in 100, or such other maximum inclination as is decided to be adopted. At each change of gradient a vertical black line is drawn in between the datum line and the surface level, and the heights of the rail level scaled off the section are figured upon these vertical lines. The horizontal distance is then scaled, and the inclination of the gradient calculated. If it is desired to make the gradient so found to rise or fall one unit in a certain multiple of chains, the height of the rail level at the further end can be raised or lowered the fractional tenths and hundredths of a foot which will enable the gradient to work out to this exact number in the given distance without remainder. The levels of the rails at each chain's length, or of the "formation level," from 9 in. to 2 ft. below the level of the upper surface of the rails, is next arrived at by calculation, and figured upon the section. Thus, referring to the diagram, the difference between the heights 109·90 and 122·20 gives 12·30 ft. in a length of $34\frac{1}{2}$ chains (2,277 ft.), which is equal to a rising gradient of 1 in 185 or ·3567 ft. per chain (66 ft.). Hence, if we alternately add 0·35 and 0·36 at each chain, we shall find we obtain 122·20 at a distance of 34½ chains from the

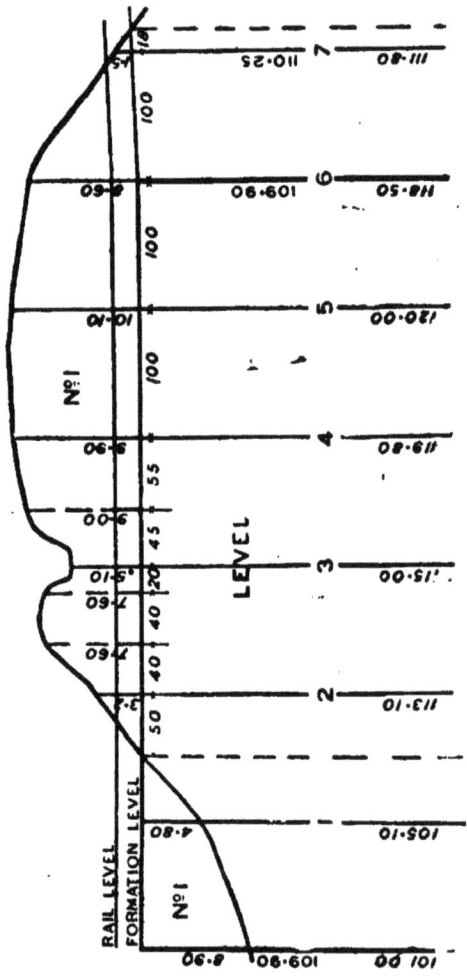

point where we measured 109·90. Thus, if either the rail level or the formation level (if preferred) be 109·90 at 6

RAILWAY WORK.

chains, it will be 109·90 plus 0·35 equals 110·25 at 7 chains; and 110·25 plus 0·36 equals 110·61 at 8 chains. The formation level is usually taken as two feet below the level of upper surface of rails. The upper row of figures upon the section shows the amount of depth of the cutting or height of the bank. The surface level at 6 chains being 118·50, by subtracting 109·90 from this we obtain 8·60 depth of cutting. The surface level at 7 chains being 111·80 and the formation level (or it may be the rail level) 110·25, the amount of cutting is here seen to be 1·55 as figured upon the section. (See pages 313, 314.)

The quantities of earthwork can be readily arrived at by the use of Bidder's tables (sold in sheets, at 2s. each or mounted 3s. 6d.). The section is headed with the number of the railway, in this case No. 1, and the banks and cuttings are each numbered in order, the banks in this section numbering 1 to 5, and the cuttings 1 to 4. The student will perceive the importance of employing a large scale for plotting the vertical measurements, especially in that portion of a longitudinal section which represents the tunnel or viaduct, and which would otherwise not so readily be appreciated by the eye. A viaduct is indicated in the place of a high embankment between 9 and 18 chains and a tunnel in the place of deep cutting between 23 and 36 chains. The adoption of a viaduct or tunnel in the place of a bank or cutting is generally a question of expense, the cheapest being usually recommended, but it is also dependent upon the geological formation and the nature of the property. All tunnels must slope to one or both of their extremities for purposes of drainage. The following diagram illustrates the application of Bidder's tables in tabulating the results. (See page 311.)

In these "Tables," published by Messrs. Vacher & Sons, of Westminster, the content in cubic yards is printed in red figures for trapezoidal solids whose length is 1 chain, width 1 foot, and whose depths at the ends from 1 to 50 feet are placed in columns so as to show the required content at the intersection of the vertical and horizontal columns. Hence in the following form as shown upon the diagram. (See pages 317, 318.)

O_1

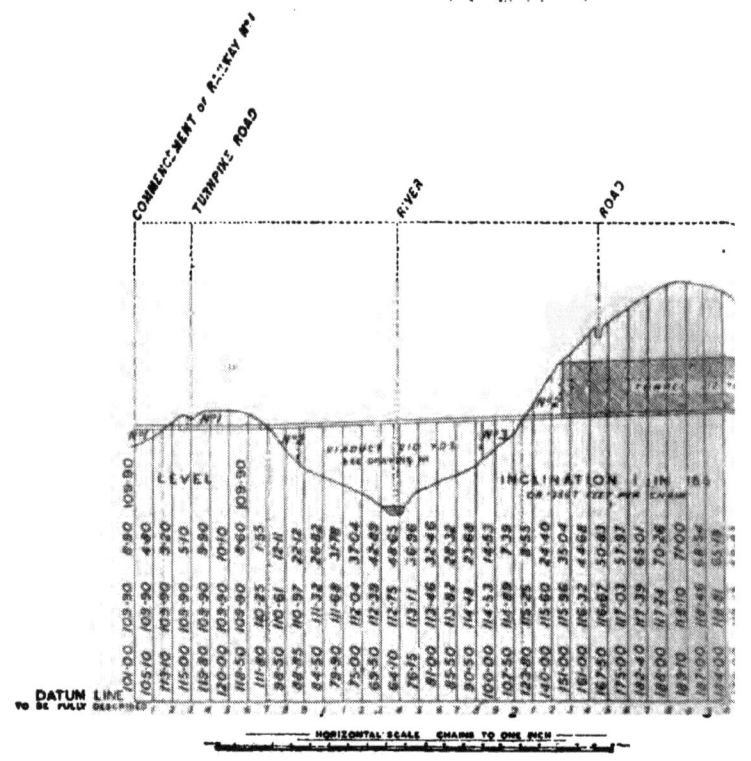

LONGITUDINAL SECTION OF RAILWAY Nº 1

BACK SET	INTER MEDIATE	FORE SET	RISE	FALL	REDUCED LEVELS	FO

Land Surveying and Levelling pages 313, 314.

No.	1 Length.	2 Heights	3 Tabular No. of Centre.	4 1 × 3.	5 Width of Centre.	6 Content of Centre, 4 × 5.

The number of the bank or cutting, taken in order from the commencement of the section is expressed for reference in the left-hand column. The length in the column headed 1 is expressed in Gunter's chains. The heights in column 2 are in feet, and the red figures found by the table are placed in column 3. In reading off the contents from the tables, the greatest height is taken in the bottom line and the less height in the left-hand column. As only feet are indicated for the heights, the nearest exact number of feet must be selected from the tables. Thus, if the heights respectively are 0·00 and 3·20 we should look for 3 in the bottom line and 0 in the left-hand vertical column. The red figure at the intersection of 0 and 3 reads 3·7 which is entered in column 3. Again, if the heights read respectively 3·20 and 7·60 we should look for 8 (as being nearer than 7 to 7·60) in the bottom line and for 3 in the left-hand vertical column, and we should find the red figure 13·4 to be entered in column 3. Column 4 means that the figures in column 1 are to be multiplied by the figures in column 3. Column 5 gives the width of the centre (see table upon the diagram), and column 6 shows that the figures in column 4 when multiplied by the figures in column 5 will give the content of the centre for the length taken. The black figures found immediately above the red figures at the intersection of the vertical and horizontal columns in the table show the content of 1 to 1 slopes whose heights are shown in column 2 of the form just described. These figures are entered in column 7 of the following form :—

7 Tabular No. of Slopes.	8 1 × 7	9 Ratio to Vertical.	10 Content of Slopes, 8 × 9.	Total Content. Cubic Yards. 6 × 10.

Column 8 contains the product of column 1 multiplied by column 7. Column 9 states the ratio to the vertical, generally taken in banks as $1\frac{1}{2}$ to 1 and in cuttings as $1\frac{1}{4}$ to 1, except rock; but the actual ratio depends upon the natural slope of the earths found upon the line of section. Column 10 gives the content of the slopes found by multiplying the figures in column 8 by the figures in column 9, and the last right hand column gives the total content in solid yards found by adding together the figures found in columns 6 and 10.

Should the heights in column 2 exceed those in the table, the contents may be found by entering with one-half, one-third, or one-fourth, &c., of the measured heights. Thus, when they fall between 50 and 100 divide them by 2, and entering the table with the quotients, multiply the first content in the table by 4 and the second by 2. When they fall between 100 and 150 divide them by 3 before the table is entered, and multiply the first content in that case by 9 and the second by 3.

The tables are calculated from the following formulæ:—
Let x = the number of feet which shows the greater height at the bottom of the table,
And y = the number of feet which shows the less height in the left hand vertical column of the table
Then first the content (black figures) column 7 = $\frac{11}{27}[(x+y)^2 - xy]$, which is based upon the ordinary formula for the volume of a truncated pyramid with parallel bases.
Second content (red figures) column 3 = $\frac{11}{9}(x+y)$
if $x = 8$ and $y = 3$
$\frac{11}{27}[(8+3)^2 - 24] = \frac{11 \times 97}{27} = 79$ (black figures)
and $\frac{11}{9}(8+3) = \frac{121}{9} = 13\cdot4$ (red figures).

In column 5 a width of 30 feet has been assumed in the case of a cutting. In the case of a bank a width of 27 feet is generally assumed. When representing, upon the plan, the widths required for the formation of the railway, it is usual to include an addition of 10 feet to the transverse width at the toe of the bank (formation level) in the case of both banks and cuttings. This gives the boundary of the area required for the purpose of valuation

RAILWAY Nº 1							
Nº	1 LENGTH IN CHAINS	2 HEIGHTS IN FEET	3 TABULAR Nº OF CENTRE RED FIGURES	4 1 × 3	5 WIDTH OF CENTRE SEE TABLE BELOW	6 CONTENT OF CENTRE 4 × 5	TAB SL 8 FI
Nº 1	0·50	0·00 3·20	3·7	1·85	30	55·50	
	0·40	3·20 7·60	13·4	5·36	,,	160·80	
	0·40	7·60 7·60	19·6	7·64	,,	235·20	
	0·20	7·60 5·10	15·9	3·18	,,	95·40	
	0·45	5·10 9·00	17·1	7·69	,,	230·70	
	0·55	9·00 9·90	23·2	12·76	,,	382·80	
	1·00	9·90 10·10	24·4	24·40	,,	722·00	
	1·00	10·10 8·60	23·2	23·20	,,	696·00	
	1·00	8·60 1·50	13·4	13·40	,,	402·00	
	0·18	1·50 0·00	2·4	·43	,,	·12·90	

Land Surveying and Levelling, pages 317, 318.

7 ⟶ULAR №OF ⟶LOPES BLACK ⟶ICURES	8 1×7	9 RATIO TO VERTICAL	10 CONTENT OF SLOPES 8×9	CUTTINGS TOTAL CONTENT CUBIC YARDS 6+10
7	3.50	1½	4.37	59.87
79	31.60	,,	39.50	200.30
156	62.40	,,	78.00	313.20
105	21.00	,,	26.25	121.65
123	55.35	,,	69.18	299.88
221	121.55	,,	151.93	534.73
244	244.00	,,	305.00	1027.00
221	221.00	,,	276.25	972.25
84	84.00	,,	105.00	507.00
3	.54	,,	.67	13.57
				4049.45

SLOPE OF BANKS 1½ TO 1
SLOPE OF CUTTINGS 1½ TO 1
EXCEPT ROCK

for the estimate of expense. In some cases a recess of 7 feet upon either side of the centre line is considered necessary, giving in the case of a cutting half of 30 feet plus 7 feet recess, or 22 feet upon each side of the centre line constant width, plus width for slope derived from heights furnished by the section.

In calculating earthwork quantities the following table of natural slopes of earths, giving a table of angles made with horizontal line, is useful for reference :—

Gravel, average	40°
Dry Sand	38°
Sand	22°
Vegetable Earth	28°
Compact Earth	50°
Shingle	39°
Rubble	45°
Clay, well drained	45°
Ditto, wet	16°

A form of level book sometimes employed in railway work is also shown under the diagram illustrating the section, in which book separate columns are provided for the heights of formation levels, embankments, and depths of excavations.

Width of Centre.

Occupation Roads	16 ft.
Single Line Railway	18 ft.
Ditto ,, ,,	20 ft.
Public Road	28 ft.
Double Line Railway	30 ft.
Ditto ,, ,,	33 ft.
Turnpike Road	38 ft.

With the use of metrical earthwork tables, the lengths in column 1 are expressed in chains of 20 metres and the heights in column 2 are expressed in metres. The width of centre in column 5 is also to be stated in metres, and the total content will be given in cubic metres. It is useful to note that $30\frac{1}{2}$ centimetres equal very nearly one foot (305 millimetres = 12·00785 inches) and that one cubic metre contains 1·3069 cubic yards. Tables for finding the contents of earthworks, when the metre is the unit of measure employed, have been compiled by Messrs.

urveying and Levelling, pp. 320, 321, 322.

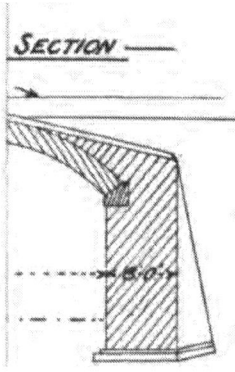

TITIES FOR PUBLIC ROAD BRIDGE

cavation	161 Cube Yards
ckwork in utments }	593 . .
hing (2ft thick)	92 Sqr Yards.
apets	37 . .
dle	17 Cube .
alte, ¾ Thick	118 Sqr .
lar or e brick }	631 Cube Feet.

RAILWAY WORK.

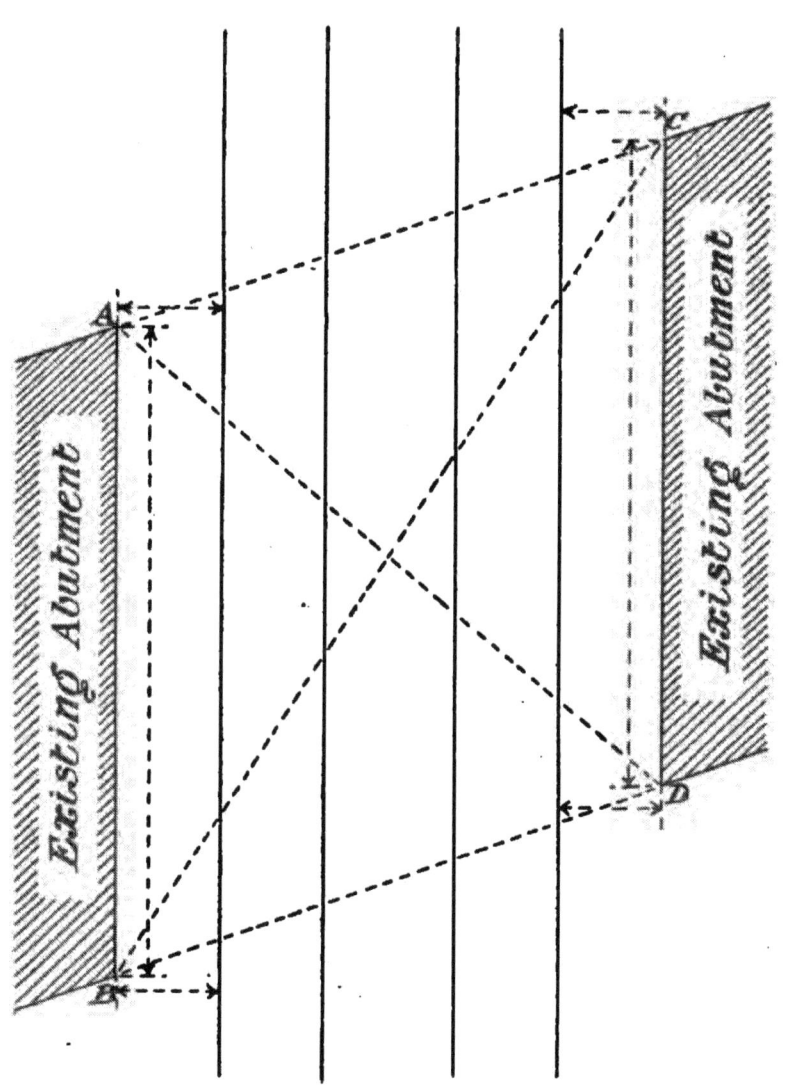

Greenbank and Piggott, arranged according to the form of Bidder's Tables. The heights given in the bottom and the vertical side columns range from zero to 16 metres, the values being expressed for every increase of one-fourth of a metre.

For application to a Parliamentary section or for an approximate estimate of the value of a railway contract, Bidder's earthwork tables are invaluable; but where greater accuracy is needed, the contents must be derived from the vertical sectional areas calculated from actual transverse sections taken at intervals along the longitudinal section.

The terms "under bridge" and "over bridge" apply to the railway. In the measurement for a skew bridge where abutments exist, it is necessary to take six main dimensions as shown by dotted lines in the annexed figures A to B, B to D, C to D, C to A, B to C and A to D. Also to record the measurement of the outside angles of the abutment from the running edge of the nearest rail. This is measured at each of the four corners in case the rails are not found parallel to the face of the abutments. The usual distance is 4 ft. 9 in. The minimum headway for an over bridge is usually 14 ft. 3 in. in the centre, and 10 ft. 3 in. at the springing or abutment side. The measurements indicated in the diagram should be taken with a steel tape.

A form for an under bridge of 25-feet span with the usual quantities is given on pages 320–322, and the height of the railway over or depth under the surface thereof must not be less than that indicated in accordance with the Standing Orders of Parliament upon the deposited section. In the case of an over bridge, the railway authorities are supposed to know their own requirements as to clearance, and usually adopt the clearance measurements named above

SCHEDULE

SHEWING THE

PROFESSIONAL PRACTICE AND CHARGES OF ARCHITECTS

(*inserted by permission of the Council of the Royal Institute of British Architects*).

1. *The usual remuneration is at 5 per cent. commission besides expenses.*—The usual remuneration for an Architect's services, except as hereinafter mentioned, is a commission of 5 per cent. on the total cost of the works executed from his designs; besides which all travelling and other incidental expenses incurred by the Architect are paid by the Employer, who may be also charged for time occupied in travelling if the work be executed at a considerable or inconvenient distance, or if more than ordinary personal attendance is required.

2. *Except for decorative work.*—But for all works in which the expenditure is mainly for skilled labour and not for materials, *e.g.*, in designs for the fittings and furniture of buildings, for their decoration with painting or mosaic, for their sculpture, for stained glass, and other like works, the Architect's charge is not made by way of commission on the cost, but should be regulated by special circumstances and conditions.

3. *Repetition in some cases justifies a lower rate.*—When several similar but distinct buildings are erected at the same time from a single specification and one set of drawings and under one contract, the commission of

5 per cent. should be charged on the cost of one such building, and a modified arrangement should be made in respect of the others.

4. *And small outlay, a higher rate of percentage.*—In works of small value, say £500 in amount, 5 per cent. is not remunerative, and the charge should be by time, or by an ascending scale reaching 10 per cent. for works under £100.

5. *Commission is to be reckoned as if for new materials and a builder employed, chargeable on all executed, and $2\frac{1}{2}$ per cent. on all omitted work.*—The commission is reckoned upon the total cost of the works, valued as if executed by a Builder, and of new materials. $2\frac{1}{2}$ per cent. is charged upon any works originally included in the contract, but subsequently omitted in execution.

This is exclusive of the charge for measuring extras and omissions.

6. *An Architect always entitled to payments on account.*—The Architect is entitled during the progress of the building to payment on account at the rate of 5 per cent. on the instalments paid to the Builder, or otherwise to half the commission on the signing of the contract, or the commencement of the works, and the remainder by instalments as above.

> N.B.—The terms of payment adopted by Her Majesty's Office of Works and Public Buildings may also be taken as an equitable method of payment on account, viz. :—
>> One-third part of the commission shall be paid to the Architect immediately after the signing of the contract;
>> One-third part shall be paid to the Architect as soon as one-half of the contract sum has been paid to the Builder;
>> And the remaining one-third part shall be paid to the Architect after the final payment to the Builder.

7. *Charge for special services.*—The above charges do not cover professional services in connection with negotiations for site, in surveying it and taking levels, in making surveys and plans of buildings to be altered, in arrangements respecting party walls or right of

lights, nor services incidental to arrangements consequent upon the failure of Builders whilst carrying out work, or in cases of subsequent litigation; but all such services are charged for in addition.

8. *And material alterations of plan by time.*—If the Employer, after having agreed to a design and had the contract drawings prepared, should have material alterations made, whether before or after the contract is prepared, an extra charge should be made, unless such alterations are rendered necessary by an unreasonable excess in the Builder's tender beyond the Architect's approximate estimate.

9. *Charge for plans and specifications only half the commission, adding one-half per cent. if tenders are obtained.*—If the Architect should have drawn out the approved design complete, with plans, elevations, sections and specification, the charge is half the commission upon the estimated cost. If he should, in addition, have procured tenders in accordance with the instruction of his employer, the charge is one-half per cent. extra.

10. *Alterations to premises may be charged at higher rate.*—For works in the alteration of premises, a special charge may be made on account of the special difficulties and trouble generally involved.

11. *The commission covers sketches, plans, details, one set of tracings, and duplicate specification, instructions, superintendence, examining accounts.*—The following are the professional services included in the ordinary charge of 5 per cent. :—

The requisite preliminary sketches, drawings and specifications sufficient for an estimate and contract.

Detailed drawings and instructions for execution.

One set of tracings and duplicate specification.

General superintendence of Works (exclusive of Clerk of the Works).

Examining and passing the accounts, exclusive of measuring and making out extras and omissions.

12. *Also approximate estimate.*—No additional remuneration is due for making an approximate estimate, such as may be obtained, for instance, by cubing out the contents. If a detailed estimate be required by the Employer, an additional percentage charge may be made.
13. *The usual charge per day.*—The charge per day made by Architects depends upon their professional position, the minimum charge being three guineas per day.
14. *All payments to Architect to be from Employer.*—The above payments alluded to in this document are to be made by the Employer to the Architect, who is not to receive commission or payment of any kind from the Builder, or any tradesman, in respect of works executed under the Architect's directions.
15. *Quantities.*—When an Architect supplies Builders with quantities, on which to form tenders for executing his designs, he should do so with the concurrence of his Employer, and it is desirable, when practicable, that the Architect should be paid by him rather than by the Builder, the cost of such extra labour not being included in the commission of 5 per cent.
16. *Ownership of drawings.*—In respect of the ownership of drawings and specifications, it has hitherto been the general custom for the Architect to be paid for their use only, those documents remaining his property.

N.B.—In case of sketches for works abandoned, this custom is recognised by Her Majesty's office of Works and Public Buildings. No authoritative decision in the Courts of Law has, however, as yet been given on the subject: it is therefore desirable, for the present at least, that the Architect should have a distinct understanding with his employer on this point.

17. *Estates.*—The charge for taking a plan of an estate, laying it out, and arranging for building upon it, should be regulated by the time, skill and trouble involved.
18. For actually letting the several plots (in ordinary cases)

a sum not exceeding a whole year's ground rent may be charged.

19. For inspecting the buildings during their progress (so far as may be necessary to ensure the conditions being fulfilled) and finally certifying for lease, the charge should be a percentage not exceeding one-half per cent. up to £5,000, and above that by special arrangement.
20. All the above fees to be exclusive of travelling expenses, and time occupied in travelling, as before mentioned.
21. The charge for the above does not include the commission for preparing specification, directing, superintending and certifying the proper formation of roads, fences and other works executed at the cost of the employer, nor for putting the plans on the leases.
22. *Valuations.*—The following definite charges are recognised for valuation of property :—

    The charge throughout is 1 per cent. on the first £1,000, and half per cent. on the remainder up to £10,000. Below £1,000, and beyond £10,000, by special arrangement.

    These charges do not include travelling expenses nor attendance before juries, arbitrators, &c.
23. *Dilapidations.*—The charge for estimating dilapidations is 5 per cent. on the estimate, and in no case less than two guineas.

NOTE.—Copies of this printed schedule can be obtained at the office of the Royal Institute of British Architects, 9, Conduit Street, Hanover Square, London, W. Re-issue: 31st March, 1882. Price Sixpence.

INDEX.

A

Angles, taking 54-56, 60-62, 71-72
" plotting 73-75, 80-83, 99-101
Arches shown on plan ... 122
Arrows 12-14

B

Back sight 169
Band tapes 19
Base lines 4, 12, 50, 52, 53-64, 95, 107, 127, 128
Beam compasses, use of 99-101
" " " ... 106
Beazeley's curve tables 216, 217
Bench mark 138, 172
Board of Agriculture ... 129, 235
Board of Trade 266, 278, 281, 282, 285, 288, 296
Book of Reference ... 255-262
Boundaries 22, 27, 134, 135-137
Bridge over and under, distinction of terms ... 324
Bridge, skew measurement 323, 324
Bridges ... 270, 320-324
Building plans, preparation 123, 236

C

Cartridge drawing paper 105, 106
Chain 11
Chaining ... 11, 20, 52, 57
Chainmen 3
Clinograph 108
Clothing 3
Colours, primary and secondary 112

Colours, conventional, for various material ... 113
Common land 243, 259
Committee rooms, Houses of Parliament, wall space 303
Computing scale ... 149-153
Contours 191
Conventional signs 114-119, 138
Corey and Barczinsky's drawing apparatus... ... 110
Cornfields on plan ... 121
Cross staff, use of ... 21, 25
Current meter 227-229
Curves, setting out ... 203-225
Cutler and Edge's curve tables 271

D

Datum 163, 170, 171, 245-247, 269
Deposited plans, preparation of 234, 238
Diagrams for Parliamentary Committees 303
Distances, inaccessible ... 26, 33
Drawing board... ... 98, 102
" table... ... 98, 99-101

E

Eidograph ... 141-145, 147
Electricity supply ... 251
Equalising areas ... 154

F

Field book ... 46-52, 84, 231
Field, object of measurements 2
" chainmen and staff-holders 3

INDEX.

	PAGE
Flying levels	163-165, 166-168
Footpaths on plan	122, 240
Foresight	169

G

Gardens shown on plan	122
Gas and water undertakings	250, 281, 285
Gauging a channel	188
Goodman's planimeter	157-162
Grading a section	310
Gradient, contouring a	196
,, calculating	271
Gravel delineation	120

H

Half-mile sections	274, 275
Harden's set square and straight-edge combined	109
Hill delineation	120, 122, 199-201
Home Office Deposits	280, 289
Houses on plan	121, 122

I

Inaccessible points	26, 27
Indian ink	113, 120
Instruments, drawing	102, 103
Intermediate sight	169, 184

L

Lakes shown on plan	121
Land surveying, its origin	1
Laths, use of	58, 194, 196
Levels, taking	169-176, 177, 187, 169
Level-book	169, 173-177
,, plotting	178-179
Light railways	266
Limits of deviation	255, 265, 270
Logarithms, use of	42

M

Machine-made papers	103
Magnetic bearing	54-56, 59
Maps	90
Marine surveying	226
Marquois scales	103

	PAGE
Meadows shown on plan	121
Mensuration, use of	1
Merrett's computing scales	153
Mounted drawing paper	103, 104, 105

N

Natural slopes of earth	319
North point on plan	54-56, 63, 123

O

Obstacles in survey lines	24, 26
Offsets	17-19, 21, 22, 58
Optical square, use of	21, 25
Ordnance datum	139, 269
Ordnance maps	127-140
Ordnance Survey, origin	127, 129

P

Parallel ruler	178, 179, 184
Parchment	105
,, tracing plans upon	106
Parish deposits	280
Parliamentary surveying, 239; outdoor work for plan, 241; general list of petitions, 243, 289; notices by advertisement, 243; book of reference, 255-260; scales, 263, 270; plan, 262; section, 268; deposits Nov. 30th, 277-282; tracings of deposited plans and sections 278; Dec. 15th, notices, 282-287; Dec. 17th, and Dec. 21st, bill deposits, 287; Dec. 23 and 31, estimate and other deposits, 290-293; Standing Order examiners, 294; deposit of money, Jan. 15th, 299, 300; committee work 301-306; alterations in Bill, 306, 307.	
Pastures shown on plan	121
Patent office tracings, size of	106
Pentagraph	141
Perpendiculars, setting out	23, 24

Q

INDEX.

	PAGE
Pier or harbour schemes	254
Plan, surface measurements, 2; attainment of area, 2; plotting	90-95, 97, 98, 106
Plan, Parliamentary	245-247, 254, 262-265, 267
Plane Table, wrist support	108, 109
Planimeter	155-157
Ploughed lands shown on plan	121
Poles	4, 9
Prismatic compass, use of	59
Professional charges	325-329
Protractors	54-56, 73-80
Provisional Orders	249, 250, 282-284, 289, 290, 296
Public roads crossed by railway	270, 271

R

Railways, 265, 269; junctions, 274, 276; quantities and Estimate	309-319
Ranging a line	4-10
Ranging rods	4
Reduced level	169
Repetition of angles	62, 65, 71, 72
Reservoir surveying	198, 200
River section	185, 226, 227
Rivers shown on plan	111, 121, 122
Road plans	235-237
Rock delineation	120
Royal Institute of British Architects, schedule of professional fees	325-329

S

Sands shown on plan	121
Satellite stations	34
Scales	91, 95, 96, 97, 123, 129, 130, 270
Scio-graphic, set square	109
Sections, Longitudinal	180
,, Plotting	178-184
,, Transverse	185, 186, 187
,, Parliamentary	245-247, 268, 269, 274-277
Sewage works	237, 285

	PAGE
Slopes, delineation of	111, 112
Soundings, taking	232, 233
Standing Orders	239, 267, 268, 270, 276, 277, 294, 308
Station pointer	81, 229
,, points	10, 57, 98, 99-101
Straight edge	90, 91, 107, 108
Stang planimeter	160
Stanley, W. F.	76, 97
Subsidiary triangles	57
Subways	252, 264

T

Tangent points	265
Tapes	11, 15
Theodolite, use of	33, 59-61, 65
Tides	140
Tidal waters	281
Tie lines	31, 32, 46, 52, 98
Tinting a plan	112
Tithe maps	128, 150, 234, 235
Town surveys	59, 61, 62, 63, 65
Tramroads	252, 265
Tramways	252, 264, 281, 285, 290, 293
Transverse sections	185, 186, 187
Traverse surveying	84-89
Trees shown on plan	121
Trigonometrical surveying	34
Trigonometry, use of	35-45
Trinity high water	139
Typographical hieroglyphics	111

U

Underground works	122

V

Vellum	105
,, tracing plans upon	105, 106
Verner's Plane Table	108, 109
Vernier readings	60, 76, 83, 87, 88

W

Water channel gauging	188
Waterway	263
Width of centre allowed in railway work	319

ADVERTISEMENTS.

J. H. STEWARD'S
SURVEYING INSTRUMENTS.

HIGH CLASS.   MODERATE PRICES.

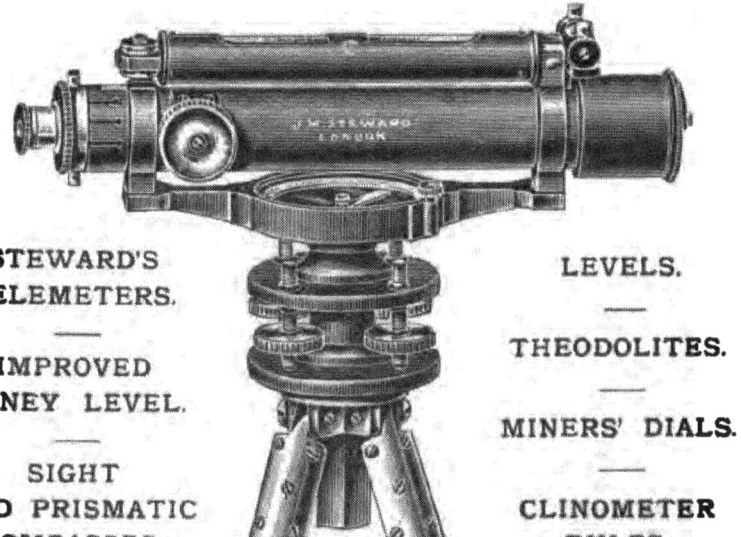

STEWARD'S TELEMETERS.

IMPROVED ABNEY LEVEL.

SIGHT AND PRISMATIC COMPASSES.

LEVELS.

THEODOLITES.

MINERS' DIALS.

CLINOMETER RULES.

ILLUSTRATED CATALOGUES
POST FREE TO ALL PARTS OF THE WORLD.

PART IV.—Surveying, Mathematical and Nautical Instruments, Drawing Instruments, Scales, Slide Rules, The R.H.S. Calculator, &c.
PART II —Meteorological Instruments, Surveying Aneroid Barometers, &c.
PART I.—Field Glasses, Telescopes, Steward's Prism Binocular, &c.

Optician to the British and Foreign Governments, Indian and Colonial Railways, &c.

406, STRAND; 457, WEST STRAND, W.C.; 7, GRACECHURCH ST., E.C., LONDON.

Telephones: { 406, Strand—Gerrard, 1,867.
{ 7, Gracechurch St.—Avenue, 930.   Telegraphic Address: "Telemeter, London."

L.S.

ADVERTISEMENTS.

W. H. HARLING,

Mathematical, Drawing, & Surveying Instrument **Manufacturer.**

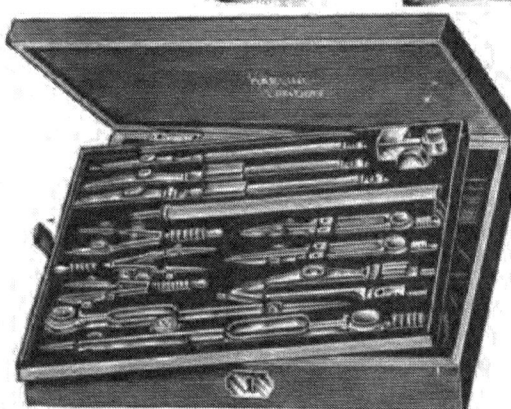

ILLUSTRATED
CATALOGUES
POST FREE.

Established 1851.

Wholesale, Retail, and Export.

Telegrams:
"Clinograph, London."

47, FINSBURY PAVEMENT, LONDON, E.C.
And, GROSVENOR WORKS, HACKNEY.

ADVERTISEMENTS.

High-class Surveying Instruments.
T. COOKE & SONS, Ltd.

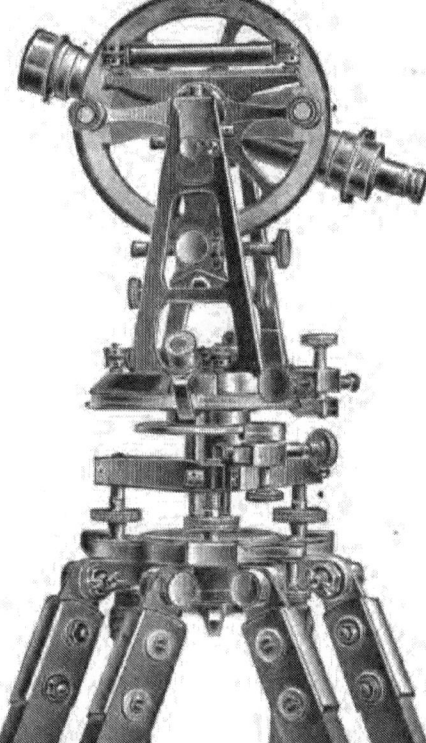

Theodolites
Tacheometers
Levels
Mining Dials
Clinometers
Plane Tables
Compasses
Sextants
Artificial Horizons
Levelling Staves

Binoculars
Telescopes
Heliographs
Chains
Steel Bands
Aneroids
Pantagraphs
Planimeters
Protractors
Drawing Instruments

Makers of Astronomical, Surveying and Drawing Instruments
TO THE
Admiralty and War Office, India Office, Colonial Office, Board of Works, Post Office, Ordnance Survey Office, Science and Art Department, South Kensington, School of Military Engineering, Chatham, and to most of the English, Indian, and Colonial Railways.

Head Office: 8, VICTORIA ST., WESTMINSTER, LONDON.
Telegrams: "COORDINATE, LONDON."

Factory: BUCKINGHAM WORKS, YORK, ENGLAND.
Telegrams: "COORDINATE, YORK."

Branch Office: 37, CASTLE ST., CAPE TOWN. | Agents: Messrs. ILBERT & CO., Shanghai.
Telegrams: "COORDINATE, CAPE TOWN." | *Telegrams:* "THEODOLITE, SHANGHAI."

"A.B.C." Code. 4th Edition. CATALOGUES SENT POST FREE ON APPLICATION.

ADVERTISEMENTS.

L. Casella,

Surveying, Mining, Meteorological and Electrical Instrument

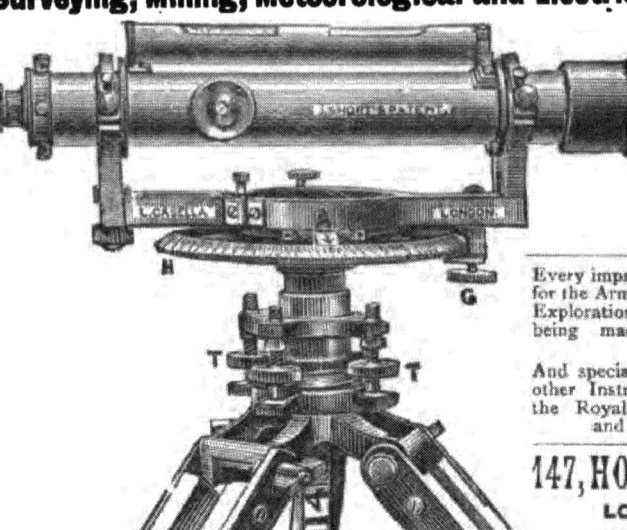

MAKER

TO THE

ADMIRALTY,
ORDNANCE,
BOARD OF TRADE,
BRITISH AND
FOREIGN
GOVERNMENTS,
VARIOUS
OBSERVATORIES,
&c., &c.

Every improvement in Instruments for the Army, Navy, and Scientific Explorations, many of these now being made in aluminium for portability.

And special sets of Portable and other Instruments, as supplied to the Royal Geographical Society and other Explorers.

147, HOLBORN BARS,
LONDON, E.C.

Telegraphic Address:
"Escutcheon, London."

Telegrams: "Dividitore, London."   Telephone: No. 1011, Holborn.

The London Drawing and Tracing Office

98, GRAY'S INN ROAD, HOLBORN, W.C.

Established 1883.   (Adjoining Holborn Town Hall.)

FOR THE PROMPT AND EFFICIENT EXECUTION OF ALL WORK USUALLY REQUIRED BY THE PROFESSION, SUCH AS

WORKING OR COMPETITION DRAWINGS
From rough sketch or description.

PERSPECTIVES
In Pen and Ink, Colour, or Monochrome.

TRACINGS
On Cloth or Tracing Paper.

PHOTO-COPIES (OR SUN PRINTS).
Brown Lines on White; Black Lines on White; Blue Lines on White; White Lines on Blue

MODELS
In Cardboard, Wood, or Plaster. Illustrations sent of Models executed.

LITHO PRINTING.
Estate Plans, Quantities, Specifications, &c., &c.

Office Hours: 9.0 to 6.0; Saturdays 9.0 to 1.0.

Cheques and Postal Orders crossed, "The London, City, and Midland Bank, Limited."

Manager: **JOHN B. THORP**, Architect and Surveyor.

ADVERTISEMENTS.

Surveyors' Institution Examinations.

COMPLETE COURSES OF PREPARATION

FOR THESE EXAMINATIONS ARE GIVEN BY

Mr. RICHARD PARRY, F.S.I., Assoc.M.Inst.C.E.
(Lecturer on Surveying at the Reading College, &c.).

The following particulars show the years in which **FIRST PLACE** has been assigned to Candidates prepared by Mr. RICHARD PARRY at the six professional examinations that were held immediately preceding the publication of this book:—

Professional Associateship Examination (for Students)
(Div. II.) 1895, 1896, 1897, 1898 and 1899.

Professional Associateship Examination (for Non-Students)
(Div. III.) 1895, 1896, 1898, 1899 and 1900.

Fellowship Examination
(Div. IV.) 1899 and 1900.

Direct Fellowship Examination
(Div. V.) 1896 and 1900.

Special Sanitary Science Examination
1900.

Mr. PARRY'S Courses of Preparation were commenced in the year 1890, and the number of successful Candidates prepared by him has increased to such an extent that at each of the three latter Examinations named above, considerably more than half of the total number of successful Candidates were his pupils.

Of the **thirty-seven** prizes awarded in the various divisions of the Professional Examinations during the above-named six years **twenty-one** were obtained by Mr. PARRY'S pupils.

Mr. PARRY also prepares Candidates for the examinations held by the Institution of Civil Engineers and the Incorporated Association of Municipal and County Engineers.

Any possible advice with respect to the Examinations or further information as to the Courses of Preparation will be given on application to

RICHARD PARRY, 82, Victoria Street, Westminster, S.W.
Telephone No. 680 Westminster.

ADVERTISEMENTS.

"The Builder" Student's Series.

Books sent from the Office, POST FREE, on receipt of remittance.

STRESSES AND STRAINS; 5/-

THEIR CALCULATION AND THAT OF THEIR RESISTANCES, BY FORMULÆ AND GRAPHIC METHODS.

By FREDERIC RICHARD FARROW, F.R.I.B.A.

With 95 Illustrations.

A Handbook for Students, particularly those preparing for the examinations of the R.I.B.A., and arranged and intended especially for those whose knowledge of mathematics is limited.

SPECIFICATIONS FOR BUILDING WORKS, 3/6

AND HOW TO WRITE THEM.

A Manual for Architectural Students.

By FREDERIC RICHARD FARROW, F.R.I.B.A.

This Manual is written with a view to meet the requirements of the Student, to show him how he should write a specification, so that when he has learnt the method and general principles, he may apply them to the particular exigencies of any building he may design.

LAND SURVEYING AND LEVELLING. 7/6

By ARTHUR T. WALMISLEY, M.I.C.E., F.S.I., F.K.C.Lond., Hon.Assoc.R.I.B.A.

With 140 Illustrations.

A useful book for a student, containing not only the usual instruction given in a Text-book on this subject, but valuable additions from the author's own experience, and an exhaustive chapter upon Parliamentary surveying.

FIELD WORK AND INSTRUMENTS. 6/-

By ARTHUR T. WALMISLEY, M.I.C.E., F.S.I., F.K.C.Lond., Hon.Assoc.R.I.B.A.

With 224 Illustrations.

In this treatise the author has endeavoured to give the Student a practical knowledge of the construction and handling of surveying instruments, and to deal with their application in the field for setting out work. The various diagrams which illustrate the subject are made as self-explanatory as possible.

ADVERTISEMENTS.

STRUCTURAL IRON AND STEEL.

By W. N. TWELVETREES, M.I.M.E.

7/-

With 234 Illustrations.

In this elementary Treatise, the author has endeavoured to present in a regular sequence some of the more important details relative to iron and steel as applied to structural work.

QUANTITIES AND QUANTITY TAKING.

By W. E. DAVIS.

3/6

A reliable Handbook for the Student pure and simple, its scope being limited to the method of procedure in the production of a good Bill of Quantities, leaving out those questions of law and other matters that do not come within the province of those for whom the work is designed. The examples given, whilst intentionally simple, to avoid confusing the Student by a mass of detail, will be found to cover almost every phase of the subject.

ARCHITECTURAL HYGIENE;

Or, SANITARY SCIENCE AS APPLIED TO BUILDINGS.

By BANISTER F. FLETCHER, A.R.I.B.A., and H. PHILLIPS FLETCHER, A.R.I.B.A., A.M.I.C.E.

5/-

With 305 Illustrations.

A concise and complete Text-book, treating the subject of Sanitary Science in all its branches (so far as it affects Architects, Surveyors, Engineers, Medical Officers of Health, Sanitary Inspectors, Plumbers, and Students generally), from the foundation of a building to its finishing and furnishing, and the application of modern methods of ventilation, lighting, and heating. It is intended to be of use to those entering for any examination in Sanitary Science.

CARPENTRY AND JOINERY:

A Text-book for Architects, Engineers, Surveyors, and Craftsmen.

By BANISTER F. FLETCHER, A.R.I.B.A., and H. PHILLIPS FLETCHER, A.R.I.B.A., AM.I.C.E.

5/-

With 424 Illustrations. 2nd Edition. Revised and corrected by the Authors.

The authors have endeavoured to meet the requirements of the craftsman, and at the same time to produce a work that will be useful to the Professional man in the designing of the various structures. They have also endeavoured to consider the desires of those who are likely to become candidates for the examination of the City and Guilds' Institute, the Carpenters' Company, and the Institute of Certified Carpenters, &c. Also for the examination in these subjects by the R.I.B.A. and the Surveyors' Institution, &c.

LONDON:

DOUGLAS FOURDRINIER, "BUILDER" Office, Catherine Street, W.C.

ADVERTISEMENTS.

Manufacturers of BLINDS for FIRE PROTECTION.
Vide British Fire Protection Committee's Report of Test, May 23, 1900.

INSIDE AND OUTSIDE **BLINDS** OF ALL KINDS.

G. A. WILLIAMS & SON,
21, QUEEN'S ROAD,
BAYSWATER, LONDON, W.

SPECIALITIES:

THE "INVISIBLE" OUTSIDE BLIND,
Unseen when not in use.

THE "UNIVERSAL" OUTSIDE BLIND,
Simplest extant.

※ WHITTAKER'S BOOKS. ※

SURVEYING AND SURVEYING INSTRUMENTS.
By G. A. T. MIDDLETON, A.R.I.B.A., M.S.A. With 41 Illustrations. 4s. 6d.
CONTENTS.—Surveys with Chain only—Obstructions in Chain-Line and Right-Angle Instruments—The Uses of the Level—Various Forms of Level and their Adjustments—The Uses of Angle-measuring Instruments—The Theodolite and other Angle-measuring Instruments—Instruments for Ascertaining Distances.

"This is a very neat little text-book, and very suitable for students preparing to pass the Institute examinations."—*Journ. of Royal Inst. of British Architects.*

PRACTICAL TRIGONOMETRY.
For the Use of Engineers, Architects, and Surveyors. By HENRY ADAMS, M.Inst.C.E., M.Inst.M.E., F.S.I., Professor of Engineering at the City of London College. 2s. 6d. net.

TABLES FOR MEASURING AND MANURING LAND.
By JOHN CULLYER. Eighteenth Edition. Cloth, 2s. 6d.

SEWAGE TREATMENT, PURIFICATION, and UTILISATION.
A Practical Manual for the Use of Corporations, Local Boards, Medical Officers of Health, Inspectors of Nuisances, Chemists, Manufacturers, Riparian Owners, Engineers, and Ratepayers. By J. W. SLATER, F.E.S. With Illustrations. 6s.

THE DRAINAGE OF HABITABLE BUILDINGS.
By W. LEE BEARDMORE, Assoc. M.Inst.C.E. Illustrated. 5s.

WHITTAKER & CO., White Hart Street, Paternoster Square, E.C.

ADVERTISEMENTS.

ASPHALTE.

The Seyssel and Metallic Lava Asphalte Co.
(Mr. H. GLENN),

Importers, Manufacturers, and Contractors

FOR

SEYSSEL AND METALLIC LAVA ASPHALTES,

SUITABLE FOR

Asylum Floors,	Dens for Animals,	Reservoirs,
Barn Floors,	Flat Roofs,	Railway Platforms,
Barrack Floors,	Foot Pavements,	Sea-Water Tanks,
Coach Houses,	Goods Sheds,	Stables,
Corn, Cotton, Hop, and Seed Floors,	Granaries,	Swimming Baths,
	Laundry Floors,	Terraces,
Corridors,	Malt Rooms,	Tun Room Floors,
Court Yards,	Milk Rooms,	Vertical to Face of Walls,
Cow Sheds,	Piggeries,	
Damp Courses,	Prison Cells,	Warehouse Floors,
Dairy Floors,	Railway Arches, Bridges, &c.	Wine Cellars,
		Waterworks.

"SEYSSEL" should be used in all cases for exposed out-door purposes.

THE METALLIC LAVA ASPHALTE

Admirably supplies the place of "Seyssel" for many in-door purposes, at a much less cost.

Prices and further particulars upon application to the

Offices: 42, Poultry, E.C., LONDON.

The Company only uses the very finest quality of materials, together with the best procurable workmanship, and undertakes to

GUARANTEE ALL WORK.

SPECIAL NOTICE.—When applying for prices for Asphalte laid complete, please state thickness, quantity, locality, and purpose for which it is to be used.

Country Builders supplied with Asphalte in bulk, with instructions for laying.

Particular attention is paid to Dairy and Tun Room Floors.

ASPHALTE CONTRACTORS TO THE FORTH BRIDGE COMPANY,
THE RIO DE JANEIRO WATERWORKS,
THE MALIGAKANDA RESERVOIR, COLOMBO, CEYLON,
and other Waterworks,
all of very considerable magnitude.

ADVERTISEMENTS.

SPRAGUE & CO., Ltd., Lithographers, Printers, &c.,

having devoted their attention for several years to the requirements of the Architectural Profession, are enabled to offer special facilities for the execution of the various descriptions of work detailed below, on moderate terms and with strict punctuality.

BILLS OF QUANTITIES, SPECIFICATIONS, REPORTS, &c. Special Writers are on the premises for these, and being constantly engaged upon them, they are thoroughly conversant with the technical terms and peculiarities of this work. Certified Copies of Bills can be supplied in a few hours if required; or, if received in the evening, by 9 o'clock next day; or by return of post to the country.

FAIR COPIES of Specifications, Contracts, Reports for Competitions, Dimensions, Accounts, &c., neatly and accurately executed.

ELEVATIONS, PLANS, AND SECTIONS FOR TENDERS. Tracings or Manuscript Copies made, or Architects' own Drawings lettered and finished.

PLANS OF ESTATES, with or without Perspective Views, tastefully lithographed and printed in colours, or highly shaded-up by hand.

PHOTO-LITHOGRAPHY rapidly executed in a superior manner.

"INK-PHOTO" process reproduces Water-Colour or Pencil Drawings or Photographs with the same facility as Pen and Ink Drawings in line. *Vide* the illustrations in "The Architect," "The Builder," &c., &c. In cases of especial emergency, any class of Drawing may be photographed and returned in course of a few hours.

PROCESS BLOCKS. We are now in a position to supply Half-tone or Line Blocks for printing with type, at short notice.

PHOTOGRAPHY. A thoroughly competent operator will be sent to take Photographs of Buildings at short notice.

CERTIFICATE BOOKS for Instalments always in stock, 3*s.*; a larger size, with Receipt, 5*s.* Bill, Dimension, and Specification Paper, Circulars, Headed Note or Letter Paper, and Memo. Forms Lithographed or Printed.

DRAWING PAPERS, Tracing Papers, &c., Office Stationery, and Sundries of every kind.

4 & 5, EAST HARDING STREET, FETTER LANE, LONDON, E.C.

ADVERTISEMENTS.

LOVE & WYMAN,
LIMITED,

Printers, Lithographers, Commercial and Drawing Office Stationers.

Tracing Linens,
　　Tracing Papers,
　Whatman's Drawing Papers.

Cartridge and Detail Papers in continuous rolls and sheets.

Mounted Papers of every description.

Ferro-Prussiate and other Sensitised Papers for copying Tracings.

Drawing Office Sundries.

General Stationery and Account Books.

TRACINGS COPIED BY ALL PROCESSES
(Electric Light used when required).

QUANTITIES LITHOGRAPHED WITH DESPATCH.

For Prices and Samples please address—

LOVE & WYMAN. LTD.,
Great Queen Street, London, W.C.
AND
3 & 4, GREAT WINCHESTER STREET, E.C.

ADVERTISEMENTS.

C. BAKER, 244, HIGH HOLBORN, LONDON.
Telegrams: "OPTIVORUM, LONDON."
Telephone: "1427, HOLBORN."

Mathematical & Surveying Instruments

DRAWING INSTRUMENTS.   STENCIL PLATES.

The "**ESTATE**" Levels.
12-inch £7 10 0
14-inch £9 10 0

14-ft. Painted Staff ... £2 0 0
66-ft. Chain 0 6 6

TRANSIT THEODOLITES.
4-inch ... £22 10 0   5-inch ... £24 10 0   6-inch ... £26 10 0

A Large Selection of SECOND-HAND Levels, Theodolites, Transits, and Drawing Instruments by the Best Makers.

CATALOGUES POST FREE.   INSTRUMENTS REPAIRED, PURCHASED, AND EXCHANGED.

ESTABLISHED 1880.

W. HARDAKER,

Surveyors' Lithographer,

13, GRAY'S INN RD., LONDON, W.C.

QUANTITIES, &c., accurately and neatly LITHOGRAPHED at very moderate rates, and with the utmost despatch. SKILLED STAFF. Plans copied, enlarged, reduced, coloured.

PHOTO PRINTS

from tracings, by the ELECTRIC LIGHT, at short notice and low prices. Architects and Surveyors assisted with Competent Draughtsmen.

QUANTITIES BOXES, 14 in. by 9 in. by 5 in., 2s. 6d. each.
6 for 12s. (or to order).

Lightning Source UK Ltd.
Milton Keynes UK
UKHW020956140819
347957UK00006B/875/P